Springer Theses

Recognizing Outstanding Ph.D. Research

For further volumes:
http://www.springer.com/series/8790

Aims and Scope

The series "Springer Theses" brings together a selection of the very best Ph.D. theses from around the world and across the physical sciences. Nominated and endorsed by two recognized specialists, each published volume has been selected for its scientific excellence and the high impact of its contents for the pertinent field of research. For greater accessibility to non-specialists, the published versions include an extended introduction, as well as a foreword by the student's supervisor explaining the special relevance of the work for the field. As a whole, the series will provide a valuable resource both for newcomers to the research fields described, and for other scientists seeking detailed background information on special questions. Finally, it provides an accredited documentation of the valuable contributions made by today's younger generation of scientists.

Theses are accepted into the series by invited nomination only and must fulfill all of the following criteria

- They must be written in good English.
- The topic should fall within the confines of Chemistry, Physics, Earth Sciences and related interdisciplinary fields such as Materials, Nanoscience, Chemical Engineering, Complex Systems and Biophysics.
- The work reported in the thesis must represent a significant scientific advance.
- If the thesis includes previously published material, permission to reproduce this must be gained from the respective copyright holder.
- They must have been examined and passed during the 12 months prior to nomination.
- Each thesis should include a foreword by the supervisor outlining the significance of its content.
- The theses should have a clearly defined structure including an introduction accessible to scientists not expert in that particular field.

Behrouz Touri

Product of Random Stochastic Matrices and Distributed Averaging

Doctoral Thesis accepted by the
University of Illinois at Urbana-Champaign, IL, USA

 Springer

Author
Dr. Behrouz Touri
University of Illinois
  at Urbana-Champaign
Urbana, IL
USA

Supervisor
Prof. Angelia Nedić
University of Illinois
  at Urbana-Champaign
Urbana, IL
USA

ISSN 2190-5053
ISBN 978-3-642-44465-4
DOI 10.1007/978-3-642-28003-0
Springer Heidelberg New York Dordrecht London

e-ISSN 2190-5061
ISBN 978-3-642-28003-0 (eBook)

*To Mehrangiz, Shahraz, Rouzbeh,
and Ronak ...*

Supervisor's Foreword

Complex networked systems are attracting a lot of attention in various scientific communities due to the rapid technological advances. Modern technological systems including wired and wireless communication systems (such as computers and mobile cells), transportation systems, power generation and distribution systems, data systems are large and distributed, both spatially and informationally. The dynamics of these complex systems is multidimensional as it includes information state dynamics, variable number of users, and time-varying connectivity structure due to mobility of the system components.

Bio-inspired models of complex networked systems have gained popularity across vast areas of research including:

- Dynamic systems and optimal control (for dealing with problems arising in flight formation, collision avoidance, and coordination of network of robots, among others);
- Signal processing (for distributed in network estimation and tracking);
- Optimization and Learning (for distributed multi-agent optimization over network, distributed resource allocation, distributed learning, and data-mining);
- Social/economical behavior (for studying social networks, spread of influence, opinion formation and opinion dynamics).

At the heart of the bio-inspired models is the model that describes how the information diffuses over the network in the absence of a central entity. In particular, the model prescribes how each agent in the system aggregates the information that the agent receives from its neighbors. In different research areas and applications, this model has been used under different names including consensus protocol, agreement/ disagreement process, information diffusion model, spread of influence, etc. The most commonly used aggregation model is an *averaging model* as it is simple for implementation, it has a minimal communication requirement, and it is robust to computational errors and noisy communications.

Over the past decade, the averaging model has been extensively studied since its dynamics is critical for understanding more complex network behavior. Most of the existing literature has explored the approaches that are based on matrix theory to investigate the stability and other aspects of averaging models. Only few attempts are made to study averaging models from dynamic systems point of view, perhaps due to the difficulties associated with a construction of a good Lyapunov function. Furthermore, most of the existing work has been in the domain of deterministic averaging models and only few recent studies have investigated random averaging models, mainly dealing with independent identically distributed models.

This monograph presents a comprehensive study of the averaging dynamics from both points of view, namely, non-negative matrix theory and dynamical system theory. The approach has many novel aspects of which the most significant are including: (1) The use of information flow over cuts in the induced graphs to measure the strength of the network diffusion process as opposed to edge-wise measure typically adopted in matrix theory approach; (2) The description of an adjoint dynamics associated with the original averaging which has not been previously used; and (3) The construction providing a whole class of Lyapunov comparison functions, which are critical in stability analysis of averaging dynamics; (4) The discovery of some key network properties that are critical for the stability of the averaging dynamics, but have not been observed before, such as the infinite flow property, feedback properties, and the balancedness of the network. These concepts are developed for both deterministic and random averaging dynamics, and used to demonstrate stability properties, to characterize the equilibrium groups and equilibrium points, as well as the convergence time of the dynamics. The results developed in this work are significantly enhancing our understanding of the averaging dynamics in complex networks, which has a potential to impact many application areas. Furthermore, the development of this monograph also extends its contributions to the domain of dynamic systems and non-negative matrix theories, as many of the results are new even when restricted to the domains of these theories, such as novel necessary and sufficient conditions for the ergodicity of the products of random row-stochastic matrices.

The monograph delivers the most concise and amazingly simple development explaining the complex behavior of averaging dynamics in distributed networked systems, which is comprehensively and rigorously exposed. However, it does not end the development but rather provides a fundamentally new look and mathematical foundations for further explorations of complex networked systems.

Urbana, November 2011   Angelia Nedić

Acknowledgments

First, I would like to express my deep gratitude to my thesis adviser, Professor Angelia Nedić, for the true research spirit that she showed to me, her guidance, her patience as well as the freedom she gave me during the research process of the current study. In addition, my extended appreciation goes towards her constant support, and her confidence in my scholarly abilities, and other numerous reasons which cannot be expressed in the space provided.

Furthermore, I would like to thank my thesis committee members, Professor Tamer Başar, Professor Geir E. Dullerud, Professor P. R. Kumar, and Professor Dušan M. Stipanović, for kindly accepting to be the members of my thesis committee and their helpful comments during the research process. I am especially thankful to Professor Başar, for the content of Chap. 7 which is done based on his fabulous ECE 580 course and also, Professor Dullerud for my knowledge on Non-Linear Control Theory which had a great impact on the contents of this thesis. I would also like to thank Professor Sekhar Tatikonda for bringing to my attention the connection between some of the results in this thesis and non-negative matrix theory.

I would also like to express my sincere gratitude to my former teachers. In particular, I would like to thank Professor Renming Song for my knowledge on Probability Theory, Professor Sean P. Meyn for teaching me Markov Chain Techniques on General State Spaces, and Mr. Bahman Honarmandian who changed my viewpoint on Mathematics when I was a high school student.

Also, I would like to thank many family members and friends who made my life in Urbana-Champaign enjoyable and sociable, especially, Rouzbeh, Ellie, and Ryan who have been a great source of energy, love, and support from the very first moment I stepped in the United States and my roommate, Amir Nayyeri, for all I learned from him and all the couch-sitting sessions! In addition, I would like to thank my friends Rasoul Etesami for proofreading this thesis, and Kunal Srivastava for his many helpful comments on my research throughout these years.

At last, but not least, I would like to dedicate this thesis to my family: Mehrangiz for her love, devotion, and dedication, Shahraz for his support and love,

Rouzbeh for being such a great role model for me throughout my life, and Ronak for her never lasting kindness.

This research was supported by the National Science Foundation under CAREER grant CMMI 07-42538. They are very much appreciated.

Contents

1 Introduction 1
   1.1 Past Work 4
   1.2 Overview and Contributions 5
   1.3 Notation 7
      1.3.1 Sets, Vectors and Matrices 7
      1.3.2 Probability Theory 9
      1.3.3 Graph Theory 10
      1.3.4 Control Theory 10
   References 11

2 Products of Stochastic Matrices and Averaging Dynamics 15
   2.1 Averaging Dynamics: Two Viewpoints 15
      2.1.1 Left Product of Stochastic Matrices 15
      2.1.2 Dynamic System Viewpoint 17
      2.1.3 Uniformly Bounded and B-Connected Chains 19
      2.1.4 Birkhoff-von Neumann Theorem 20
   References 21

3 Ergodicity of Random Chains 23
   3.1 Random Weighted Averaging Dynamics 23
   3.2 Ergodicity and Infinite Flow Property 24
   3.3 Infinite Flow Graph and ℓ_1-Approximation 28

4 Infinite Flow Stability 35
   4.1 Infinite Flow Stability 35
   4.2 Feedback Properties 37
   4.3 Comparison Functions for Averaging Dynamics 40
      4.3.1 Absolute Probability Process 40
      4.3.2 A Family of Comparison Functions 42
      4.3.3 Quadratic Comparison Functions 45

	4.3.4 An Essential Relation	47
4.4	Class $\mathcal{P}^*$	49
4.5	Balanced Chains	55
	4.5.1 Absolute Probability Sequence for Balanced Chains	58
References		62

5 Implications ... 65
- 5.1 Independent Random Chains ... 65
  - 5.1.1 Rate of Convergence ... 67
  - 5.1.2 i.i.d. Chains ... 73
- 5.2 Non-Negative Matrix Theory ... 76
- 5.3 Convergence Rate for Uniformly Bounded Chains ... 78
- 5.4 Link Failure Models ... 80
- 5.5 Hegselmann-Krause Model for Opinion Dynamics ... 82
  - 5.5.1 Hegselmann-Krause Model ... 82
  - 5.5.2 Loose Bound for Termination Time of Hegselmann-Krause Dynamics ... 83
  - 5.5.3 An Improved Bound for Termination Time of Hegselmann-Krause Model ... 85
- 5.6 Alternative Proof of Borel-Cantelli lemma ... 91
- References ... 92

6 Absolute Infinite Flow Property ... 93
- 6.1 Absolute Infinite Flow ... 93
- 6.2 Necessity of Absolute Infinite Flow for Ergodicity ... 97
  - 6.2.1 Rotational Transformation ... 97
  - 6.2.2 Necessity of Absolute Infinite Flow Property ... 101
- 6.3 Decomposable Stochastic Chains ... 101
- 6.4 Doubly Stochastic Chains ... 105
  - 6.4.1 Rate of Convergence ... 107
  - 6.4.2 Doubly Stochastic Chains Without Absolute Infinite Flow Property ... 109
- References ... 112

7 Averaging Dynamics in General State Spaces ... 113
- 7.1 Framework ... 113
- 7.2 Modes of Ergodicity ... 115
- 7.3 Infinite Flow Property in General State Spaces ... 118
- 7.4 Quadratic Comparison Function ... 121
- References ... 126

8 Conclusion and Suggestions for Future Works ... 127
- 8.1 Conclusion ... 127
- 8.2 Suggestions for Future Works ... 128

Appendix A: Background Material of Probability Theory 131

Appendix B: Background Material on Real Analysis and Topology... 137

Index 141

Chapter 1
Introduction

In this thesis, we study infinite product of deterministic and random stochastic matrices. This mathematical object is one of the main analytical tools that is frequently used in various problems including: distributed computation, distributed optimization, distributed estimation, and distributed coordination. In many of these problems, a common goal is attempted to be achieved among a set of agents while there is no central coordination among them. A common theme for solving those problems is to reach a form of agreement by performing *distributed averaging* among the agents, which leads to an alternative way for the study of product of (deterministic) stochastic matrices, i.e. the study of *weighted averaging dynamics*. A weighted averaging dynamics is a dynamics of the form:

$$x(k + 1) = A(k)x(k) \quad \text{for } k \geq 0, \tag{1.1}$$

where $A(k)$ is a (row) stochastic matrix for any $k \geq 0$ and $x(0) \in \mathbb{R}^m$ is arbitrary. Some motivational applications for such a study are discussed in the sequel:

Distributed Optimization

Consider a network of m agents such that each of them has a private convex objective function $f_i(x)$ which is defined on $\mathbb{R}^n$ for some $n \geq 1$. Suppose that we want to design an algorithm that solves the following optimization problem:

$$\begin{aligned} \text{minimize:} \quad & \textstyle\sum_{i=1}^{m} f_i(x) \\ \text{subject to:} \quad & x \in \mathbb{R}^n. \end{aligned} \tag{1.2}$$

The goal is to solve the problem (1.2) distributively over the network by limited local coordination of agents' actions. This problem has been studied in [1–6] both in deterministic time-changing networks and random i.i.d. networks. A generalization of the problem (1.2) is studied in [7] where each agent has a convex constraint set $C_i \subset \mathbb{R}^n$ and the goal is to solve the problem over the intersection set $\bigcap_{i=1}^{m} C_i$.

Optimization problem (1.2) can be solved by using the following scheme: suppose that at time k, agent i's estimate of a solution to (1.2) (which is assumed to exist) is $x^i(k)$. Then, we set

B. Touri, *Product of Random Stochastic Matrices and Distributed Averaging*,
Springer Theses, DOI: 10.1007/978-3-642-28003-0_1,
© Springer-Verlag Berlin Heidelberg 2012

$$x^i(k+1) = \sum_{j=1}^{m} a_{ij}(k)x^j(k) - \alpha^i(k)d_i(k). \tag{1.3}$$

In Eq. (1.3), $A(k) = \{a_{ij}(k)\}_{i,j\in[m]}$ is a doubly stochastic matrix. The chain $\{A(k)\}$ is assumed to possess certain properties. The variable $\alpha^i(k)$ is the stepsize of the ith agent at time k which also satisfies certain conditions, and $d_i(k)$ is a subgradient vector of the function $f_i(x)$ at $x^i(k)$.

If in Eq. (1.3), we have $d_i(k) = 0$, i.e., $f_i(x)$ is constant for all $i \in [m]$, the dynamics (1.3) reduces to the dynamics (1.1) which is the focal point of the current study. Nevertheless, for the general case of non-trivial objective functions, convergence analysis of the algorithm in Eq. (1.3) relies on the stability properties of the dynamic system (1.1) driven by the chain $\{A(k)\}$.

Synchronization

Consider a network with m processors. Each of the m processors can compute its local time τ_i using its own Central Processing Unit (CPU) clock. Ideally, after the calibration, each processor's local time should be equal to the Coordinated Universal Time t. However, due to the hardware imperfections of CPU clocks, different processors, even if they share the same hardware architecture, might have different time stamps for a certain event. A first order model to describe such a drift is the following linear model:

$$\tau_i(t) = a_i t + b_i,$$

where $\tau_i(t)$ is the clock reading of the ith processor at time t, while a_i and b_i are the ith processor's clock skew and clock offset, respectively. Ideally, we should have $a_i = 1$ and $b_i = 0$ for all $i \in \{1, \ldots, m\}$. However, this is not the case in many real situations. In some applications, inaccurate and asynchronized time readings might not cause any problem. However, in certain applications, such as multiple target tracking scheme [8], time synchronization of different processors is crucial.

In [9, 10], a clock synchronization scheme has been proposed and developed based on the convergence of dynamic system (1.1). A similar approach has been proposed in [11] for clock synchronization in sensor networks. The main idea in those works is to mix the clock readings τ_i's through the underlying network and *align* each local time with a virtual universal clock $\tau_u(t) = a_u t + b_u$. The proposed alignment schemes use the convergence properties of the dynamics (1.1) under certain connectivity conditions on the underlying communication network of the m processors.

Robotics

Study of the dynamic system (1.1) has various applications in networks of robots, especially when there is no central coordination among the robots. An example of those applications is achieving rendezvous in a network of robots. To describe the problem, consider a network of m robots. Suppose that the ith robot is initially positioned at $x_i(0) \in \mathbb{R}^2$, where $i \in [m]$ and $[m] = \{1, \ldots, m\}$. The goal is to

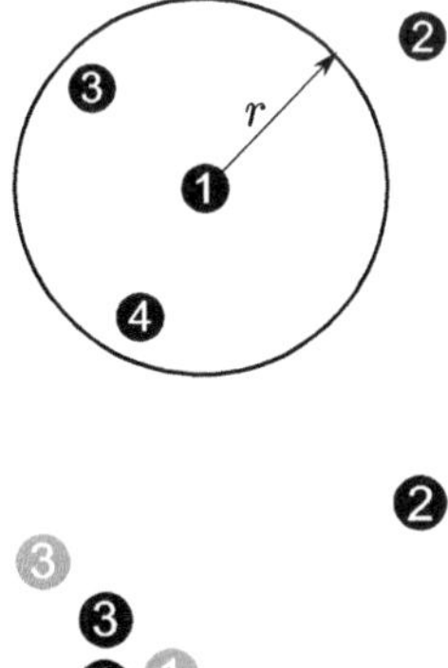

Fig. 1.1 At each time, every robot observes the positions of the robots at its r-distance. In this configuration, robot 1 observes the positions of the robots 1 (itself), 3, and 4

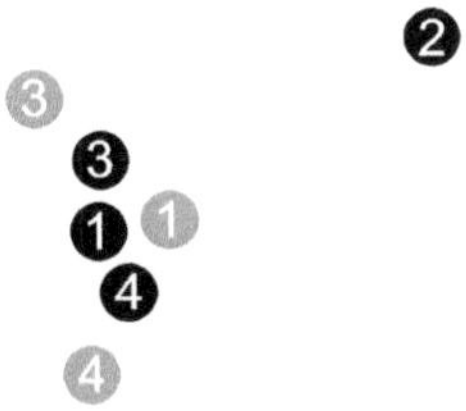

Fig. 1.2 The positions of the 4 robots in Fig. 1.1 after one iteration of the distributed rendezvous algorithm

gather the robots at a common point, a *rendezvous point*. For rendezvous, consider the following recursive algorithm:

(1) At time $k \geq 0$, robot $i \in [m]$ measures or receives the positions of all of the robots at a distance r, i.e., the positions of the robots in the set $\mathcal{N}_i(k) = \{j \in [m] \mid \|x_i(k) - x_j(k)\| \leq r\}$ (see Fig. 1.1).

(2) At time $k \geq 0$, robot $i \in [m]$ computes the average position of the neighboring robots $\bar{x}_i(k) = \frac{1}{|\mathcal{N}_i(t)|} \sum_{j \in |\mathcal{N}_i(k)|} x_j(k)$, where $|\mathcal{N}_i(k)|$ is the number of elements in $\mathcal{N}_i(k)$.

(3) Robot i moves to the point $\bar{x}_i(k)$ before the communication time $k + 1$, i.e., $x_i(k + 1) = \bar{x}_i(k)$.

One iteration of the above algorithm is illustrated in Fig. 1.2. This algorithm is motivated by the works in [12, 13] on modeling of social opinion dynamics and it is known as Hegselmann–Krause model [14]. We study this model more extensively in Chap. 5 .

Observe that in the Hegselmann–Krause algorithm, the evolution of the position vector $x(k) = (x_1(k), \ldots, x_m(k))^T$ follows the dynamics (1.1). In fact, rendezvous resulting from the Hegselmann-Krause algorithm is equivalent to achieving consensus in dynamics (1.1), a concept that will be introduced later in Chap. 2.

Theoretical Motivation

One of the main motivations of this study is the desire to extend the following well-known and widely used result for ergodic Markov chains.

Lemma 1.1 *Let A be an irreducible and aperiodic matrix. Then A^k converges to a rank one matrix as k approaches infinity.*

In Chaps. 4 and 5, we prove a generalization of this result for not only the product of time-varying stochastic matrices but also the product of independent random stochastic matrices. We show that, in fact, many seemingly different results in the field of consensus and distributed averaging are just special cases of this general result.

1.1 Past Work

Unfortunately, the diversity and numerous publications in this area makes it almost impossible to have an extensive and thorough literature review on this field. Here, we review (relatively) few of the previous work on the study of product of random and deterministic sequences of stochastic matrices by focusing mainly on the literature that shaped this thesis.

The study of forward product of an inhomogeneous chain of stochastic matrices is closely related to the limiting behavior, especially ergodicity, of inhomogeneous Markov chains. One of the earliest studies on the forward product of inhomogeneous chains of stochastic matrices is the work of Hajnal in [15]. Motivated by a homogeneous Markov chain, Hajnal formulated the concepts of ergodicity in weak and strong senses for inhomogeneous Markov chains, and developed some sufficient conditions for both weak and strong ergodicity of such chains. Using the properties of scrambling matrices that were introduced in [15], Wolfowitz [16] gave a condition under which all the chains driven from a finite set of stochastic matrices are strongly ergodic. In his elegant work [17], Shen gave geometric interpretations and provided some generalizations of the results in [15] by considering vector norms other than $\| \cdot \|_\infty$, which was originally used in [15] to measure the scrambleness of a matrix.

One of the notable works, which is used in some of the main results of this thesis is the elegant manuscript of Kolmogorov [18]. There, he studied the behavior of a Markov chain that is started in $-\infty$. To study such Markov chains, he introduced the concept of an *absolute probability* sequence for a chain of stochastic matrices and proved existence of such a sequence for any Markov chain. This sequence and its existence will play a central role in the development of this thesis.

The study of backward product of row-stochastic matrices, however, is motivated by different applications all of which are in search of a form of a consensus among a set of processors, individuals, or agents. DeGroot [19] studied such a product (for a homogeneous chain) as a tool for reaching consensus on a distribution of a certain unknown parameter among a set of agents. Later, Chatterjee and Seneta [20] provided a theoretical framework for reaching consensus by studying the backward product of an inhomogeneous chain of stochastic matrices. Motivated by the theory of inhomogeneous Markov chains, they defined the concepts of weak and strong ergodicity in this context, and showed that these two properties are equivalent. Furthermore, they developed the theory of coefficients for ergodicity. Motivated by some distributed computational problems, in [2], Tsitsiklis and Bertsekas studied such a product from the dynamical system point of view. In fact, they considered a dynamics that accomodates an exogenous input as well as delays in the system. Through the study of such dynamics, they gave more practical conditions for a chain to ensure ergodicity and consensus. The work in [1, 2] had a great impact on the subsequent studies of distributed estimation and control problems. Another notable work in the area of distributed averaging is [21]. There, a similar result as in [1, 2] for consensus and ergodicity under a slightly more general condition was given. Also, other interesting questions such as existence of a quadratic Lyapunov function for averaging dynamics were raised there. Non-existence of quadratic Lyapunov functions for general averaging dynamics was verified numerically there and was analytically proven in [22]. A notable work on the study of the convergence and stability of averaging dynamics is [23], where a general condition for convergence and stability of averaging dynamics is established. In [24, 25], convergence rate and efficiency of different averaging schemes are discussed and compared.

The common ground in the study of both forward and backward products of stochastic matrices are the chains of doubly stochastic matrices. By transposing the matrices in such a chain, forward products of matrices can be transformed into backward products of the transposes of the original matrices. However, the transposition of a row-stochastic matrix is not necessarily a row-stochastic matrix, unless the matrix is doubly stochastic. Therefore, in the case of doubly stochastic matrices, any property of backward products can be naturally translated into forward products.

The study of random product of stochastic matrices dates back to the early work in [26], where the convergence of the product of i.i.d. random stochastic matrices was studied using the algebraic and topological structures of the set of stochastic matrices. This work was further extended in [27–29] by using results from ergodic theory of stationary processes and their algebraic properties. In [30], the ergodicity and consensus of the product of i.i.d. random stochastic matrices with almost sure diagonal entries were studied. The main result in [30] can be concluded from the works in [27, 28], however, the approach used there was quite different. Independently, the same problem was tackled in [31], where an exponential convergence bound was established. The work in [30] was extended to ergodic stationary processes in [32].

This thesis is also related to opinion dynamics in social networks [12, 13] and its generalizations as discussed in [23, 33, 34], consensus over random networks [35], optimization over random networks [4], and the consensus over a network with random link failures [36]. Related are also gossip and broadcast-gossip schemes giving rise to a random consensus over a given connected bi-directional communication network [37–40]. On a broader basis, this work is related to the literature on the consensus over networks with noisy links [41–44] and the deterministic consensus in decentralized systems models [1, 2, 21, 24, 45–49], including the effects of quantization and delay [40, 50–54].

1.2 Overview and Contributions

As already mentioned, this thesis is mainly devoted to the study of products of stochastic matrices and its generalizations to random chains and general state spaces. Here, we provide an overview and summarize the main contributions of the thesis.

In Chap. 3, we discuss the framework for studying a product of random stochastic matrices and its corresponding dynamics driven by such matrices. We introduce the concept of *infinite flow property* which is hidden in all the previously known results on ergodicity of deterministic chains. We show that this property is in fact *necessary* for ergodicity of any chain. Motivated by this result, we introduce the concept of *infinite flow graph*. By introducing ℓ_1 *-approximation* of stochastic chains, we show that the limiting behavior of a product of stochastic matrices is closely related to the connectivity of the *infinite flow graph* associated with such a chain.

In Chap. 4, we study the converse statements of the results developed in Chap. 3. We first introduce the concept of *infinite flow stability* for a random chain. In our attempt to specify classes of random chains that exhibit infinite flow stability, we first study a property which has also been commonly assumed in various forms in the previous studies in this field, i.e. the *feedback property*. We define different feedback properties and investigate their relations with each other. Motivated by an absolute probability sequence for a chain of stochastic matrices, we introduce the concept of *absolute probability process* for an adapted random chain. We show that using an absolute probability process, one can define infinitely many comparison functions for the corresponding adapted random chain including a quadratic one. Using the quadratic comparison function, we show that any chain in a certain class of adapted random chains with weak feedback property is infinite flow stable. Then, we define a class of *balanced chains* that includes nearly all the previously known ergodic chains. We show that any balanced chain with feedback property is in fact infinite flow stable.

Then, in Chap. 5, we study some of the implications of the results in Chaps. 3 and 4 for products of independent random stochastic matrices. We show that, under general conditions of balancedness and feedback property, the ergodic behavior of an independent random chain and its expected chain are equivalent. We also develop a rate of convergence result for a class of independent random chains. Then, we visit the problem of link-failure on random chains and develop a condition under which link failure does not affect the limiting behavior of the chain. We also discuss the Hegselmann–Krause model for opinion dynamics. We show how an application of the developed machinery results in an upper bound of $O(m^4)$ for the termination time of the Hegselmann–Krause model. Finally, we present an alternative proof of the second Borel–Cantelli lemma using the developed results.

In Chap. 6, we extend the notion of infinite flow property to *absolute infinite flow property*. We prove that this stronger property is also necessary for ergodicity of any stochastic chain. We do this through the introduction of a *rotational transformation* of a stochastic chain with respect to a permutation chain. Then, we show that limiting behavior of a stochastic chain is invariant under rotational transformation. Using this result, we prove that ergodicity of any doubly stochastic chain is equivalent to absolute infinite flow property. Also, using the rotational transformation, we show that any products of doubly stochastic matrices is essentially convergent, i.e. it is convergent up to a sequence of permutation. We also develop a rate of convergence result for doubly stochastic chains based on the rate results established in Chap. 5.

Finally, we extend the notion of averaging and product of stochastic matrices to general measurable state spaces. There, we define several modes of ergodicity and extend the notion of infinite flow property. We prove that in general state spaces, this property is necessary for the weakest form of ergodicity. We define an absolute probability sequence for a chain of stochastic kernels and we show that a chain of stochastic integral kernels with an absolute probability sequence admits infinitely many comparison functions. We also derive the decrease rate of the associated quadratic comparison function along the trajectories of any dynamics driven by such chains.

The main contributions of this thesis include:

- Developing of new concepts, such as *infinite flow property, infinite flow graph, feedback properties, mutual ergodicity, infinite flow stability, absolute probability process, balanced chains, absolute infinite flow property*, and showing their relevance to the study of product of random and deterministic stochastic matrices and averaging dynamics.
- Developing of new techniques, such as *randomization technique, ℓ_1-approximation, rotational transformation*, to study products of random and deterministic stochastic matrices.
- Establishing the existence of infinitely many comparison functions for averaging dynamics including a quadratic one.
- Developing necessary and sufficient conditions for ergodicity of deterministic and random stochastic chains.
- Extending a fundamental result for the study of irreducible and aperiodic homogeneous Markov chains to inhomogeneous chains and random chains of stochastic matrices.
- Developing a unified approach for convergence rate analysis of the averaging and consensus algorithms.
- Developing a new bound for the convergence time of the Hegselmann–Krause model for opinion dynamics.
- Formulating and study of the averaging dynamics, ergodicity, and consensus over general state spaces.

This thesis is based on the work presented in published and under-review papers, and technical reports [55–63].

1.3 Notation

The notation used in this thesis is aimed to be as intuitive as possible. One may skip this section and refer back to it, if any notation is confusing.

1.3.1 Sets, Vectors and Matrices

We use $\mathbb{R}$ and $\mathbb{Z}$ to denote the sets of real numbers and integers, respectively. Furthermore, we use $\mathbb{R}^+$ and $\mathbb{Z}^+$ to denote the sets of non-negative real numbers and non-negative integers, respectively. We use $\mathbb{R}^{m \times m}$ to denote the set of $m \times m$ real-valued matrices. We use $[m]$ to denote the set $\{1, \ldots, m\}$. For a set $S \subset [m]$, we let $|S|$ be the cardinality of the set S.

We view all vectors as column vectors. For a vector x, we write x_i to denote its ith entry, and we write $x \geq 0$ ($x > 0$) to denote that all its entries are nonnegative (positive). We use x^T to denote the transpose of a vector x. We write $\|x\|$ to denote the standard Euclidean vector norm i.e., $\|x\| = \sqrt{\sum_i x_i^2}$ and we write $\|x\|_p = (\sum_{i=1}^m |x_i|^p)^{1/p}$ to denote the p-norm of x where $p \in [1, \infty]$. We use e_i to denote the vector with the ith entry equal to 1 and all other entries equal to 0, and we write e for the vector with all entries equal to 1.

For a given set C and a subset S of C, we write $S \subset C$ to denote that S is a proper subset of C. A proper and non-empty subset S of a set C is said to be a *nontrivial* subset of C. We write $\bar{S}$ to denote the complement of the set $S \subseteq C$, i.e. $\bar{S} = \{\alpha \in C \mid \alpha \notin S\}$. We denote the power set of a set C, i.e. the set of all subsets of C, by $\mathscr{P}(C)$.

We denote the identity matrix by I and the matrix with all entries equal to one by J. For a matrix A, we use A_{ij} or $[A]_{ij}$ to denote its (i, j)th entry, A_i and A^j to denote its ith row and jth column vectors, respectively, and A^T to denote its transpose. We write $\|A\|_p$ for the matrix p-norm induced by the vector p-norm, i.e.

$$\|A\|_p = \max_{\|x\|_p = 1} \|Ax\|_p.$$

As for the vector 2-norm, we denote $\|A\|_2$ by $\|A\|$.

A matrix A is *row-stochastic* when its entries are nonnegative and the sum of a row entries is 1 for all rows. Since we deal exclusively with row-stochastic matrices, we refer to such matrices simply as *stochastic*. We denote the set of $m \times m$ stochastic matrices by $\mathbb{S}_m$. A matrix A is *doubly stochastic* when both A and A^T are stochastic. We often refer to a matrix sequence as a *chain*. We say that a chain $\{A(k)\}$ of matrices is *static* if it does not depend on k, i.e. $A(k) = A(0)$ for all $k \geq 0$, otherwise we say that $\{A(k)\}$ is an *inhomogeneous* or *time-varying* chain. Similarly, we say that a sequence of vectors $\{\pi(k)\}$ is static if $\pi(k) = \pi(0)$ for all $k \geq 0$.

For a vector $\pi \in \mathbb{R}^m$ and a scalar $\alpha \in \mathbb{R}$, we write $\pi \geq \alpha$ ($\pi > \alpha$) if $\pi_i \geq \alpha$ ($\pi_i > \alpha$) for all $i \in [m]$. Similarly, for a matrix $A \in \mathbb{R}^{m \times m}$ we write $A \geq \alpha$ ($A > \alpha$), if $A_{ij} \geq \alpha$ ($A_{ij} > \alpha$) for all $i, j \in [m]$. Finally, for a sequence of vectors $\{\pi(k)\}$, we write $\{\pi(k)\} \geq \alpha$ if $\pi(k) \geq \alpha$ for all $k \geq 0$.

For an $m \times m$ matrix A, we use the following abbreviation:

$$\sum_{i<j} A_{ij} = \sum_{i=1}^m \sum_{j=i+1}^m A_{ij}.$$

For a vector $\pi = (\pi_1, \ldots, \pi_m)^T \in \mathbb{R}^m$, we use $\operatorname{diag}(\pi)$ to denote the diagonal matrix

$$\mathrm{diag}(\pi) = \begin{bmatrix} \pi_1 & 0 & \cdots & 0 & 0 \\ 0 & \pi_2 & \cdots & 0 & 0 \\ \vdots & \vdots & \ddots & \vdots & \vdots \\ 0 & 0 & \cdots & \pi_{m-1} & 0 \\ 0 & 0 & \cdots & 0 & \pi_m \end{bmatrix}.$$

Given a nonempty index set $S \subseteq [m]$ and a matrix A, we write A_S to denote the following summation:

$$A_S = \sum_{i \in S, j \in \bar{S}} A_{ij} + \sum_{i \in \bar{S}, j \in S} A_{ij}.$$

Note that A_S satisfies

$$A_S = \sum_{i \in S, j \in \bar{S}} (A_{ij} + A_{ji}).$$

An $m \times m$ stochastic matrix P is a *permutation matrix* if it contains exactly one entry equal to 1 in each row and each column. Given a permutation matrix P, we use $P(S)$ to denote the image of an index set $S \subseteq [m]$ under the permutation P; specifically

$$P(S) = \{i \in [m] \mid P_{ij} = 1 \text{ for some } j \in S\}.$$

We note that a set $S \subset [m]$ and its image $P(S)$ under a permutation P have the same cardinality, i.e., $|S| = |P(S)|$. Furthermore, for any permutation matrix P and any nonempty index set $S \subset [m]$, the following relation holds:

$$\sum_{i \in P(S)} e_i = P \sum_{j \in S} e_j.$$

We denote the set of $m \times m$ permutation matrices by $\mathscr{P}_m$. Since there are $m!$ permutation matrices of size m, we may assume that the set of permutation matrices is indexed by the index set $[m!]$, i.e., $\mathscr{P}_m = \{P^{(\xi)} \mid 1 \leq \xi \leq m!\}$. Also, we say that $\{P(k)\}$ is a *permutation sequence* if $P(k) \in \mathscr{P}_m$ for all $k \geq 0$. The sequence $\{I\}$ is the permutation sequence $\{P(k)\}$ with $P(k) = I$ for all k, and it is referred to as the *trivial permutation sequence*.

1.3.2 Probability Theory

Consider a probability space $(\Omega, \mathcal{F}, \mathrm{Pr}\,(\cdot))$ where Ω is a set (often referred to as the sample space), $\mathcal{F}$ is a σ-algebra on Ω, and $\mathrm{Pr}\,(\cdot)$ is a probability measure on $(\Omega, \mathcal{F})$. We refer to members of $\mathcal{F}$ as events. We denote the Borel σ-algebra on $\mathbb{R}$ by $\mathcal{B}$.

We say that a property $\mathbf{p}$ holds almost surely if the set $\{\omega \in \Omega \mid \omega \text{ does not satisfy } \mathbf{p}\}$ is an event and

$$\Pr\left(\{\omega \mid \omega \text{ does not satisfy } \mathbf{p}\}\right) = 0.$$

We use the abbreviation a.s. for almost surely. We denote the characteristic function of an event E by $\mathbf{1}_E$, i.e. $\mathbf{1}_E(\omega) = 1$ for $\omega \in E$ and $\mathbf{1}_E(\omega) = 0$, otherwise. We say that an event E is a trivial event if $\Pr(E) = 0$ or $\Pr(E) = 1$, or in other words, it is equal to the empty set or Ω, almost surely.

We denote the expected value of a random variable u by $\mathsf{E}[u]$ and the conditional expectation of u conditioned on a σ-algebra $\tilde{\mathcal{F}}$ by $\mathsf{E}\left[u \mid \tilde{\mathcal{F}}\right]$.

We say that $x : \Omega \to \mathbb{R}^m$ is a random vector if each coordinate function x_i is a random variable for all $i \in [m]$. Likewise, we say that $W : \Omega \to \mathbb{R}^{m \times m}$ is a random matrix if W_{ij} is a random variable for all $i, j \in [m]$.

For a collection Θ of subsets of Ω, we let $\sigma(\Theta)$ to be the smallest σ-algebra containing Θ. For a random variable u, we let $\sigma(u) = \sigma(\{u^{-1}(B) \mid B \in \mathcal{B}\})$.

1.3.3 Graph Theory

We view an *undirected graph* G on m vertices as an ordered pair $([m], \mathcal{E})$ where $\mathcal{E}$ is a subset of $\{\{i, j\} \mid i, j \in [m]\}$. We refer to $[m]$ as the vertex set and we refer to $\mathcal{E}$ as edge set of G. If $\{i, j\} \in \mathcal{E}$, we say that j is a neighbor of i. We denote the set of all neighbors of $i \in [m]$ by $\mathcal{N}_i(G) = \{j \in [m] \mid \{i, j\} \in \mathcal{E}\}$. We denote the set of all undirected graphs on the vertex set $[m]$ by $\mathcal{G}([m])$.

A path between two vertices $v_1 = i \in [m]$ and $v_p = j \in [m]$ in the graph $G = ([m], \mathcal{E})$ is an ordered sequence of vertices $(v_1, v_2, \ldots, v_p)$ such that $\{v_\ell, v_{\ell+1}\} \in \mathcal{E}$ for all $\ell \in [p-1]$. We say that G is a connected graph if there is a path between any distinct vertices $i, j \in [m]$. We say that $S \subseteq [m]$ is a *connected component* if there is a path between any two vertices $i, j \in S$ and S is the maximal set with this property.

We view a *directed graph* G on m vertices as an ordered pair $([m], \mathcal{E})$ where $\mathcal{E} \subseteq \{(i, j) \mid i, j \in [m]\}$. We say that $j \in [m]$ is a neighbor of i if $(i, j) \in \mathcal{E}$ and as in the case of undirected graph, we denote the set of neighbors of i in $G = ([m], \mathcal{E})$ by $\mathcal{N}_i(G)$. In this case, a directed path between a vertex $v_1 = i \in [m]$ and $v_p = j \in [m]$ in $G = ([m], \mathcal{E})$ is an ordered sequence of vertices $(v_1, v_2, \ldots, v_p)$ such that $(v_\ell, v_{\ell+1}) \in \mathcal{E}$ for all $\ell \in [p-1]$. We say that a directed graph G is *strongly connected* if there is a directed path between any two distinct vertices $i, j \in [m]$.

For an $m \times m$ non-negative matrix A, we say that $G = ([m], \mathcal{E})$ is the graph induced by the positive entries of A, or simply the graph induced by A if $\mathcal{E} = \{(i, j) \mid A_{ij} > 0\}$.

1.3.4 Control Theory

Let $f : X \times \mathbb{Z}^+ \to X$ for some space X. Let $t_0 \geq 0$ be an arbitrary non-negative integer, and let $x(t_0) \in X$. Let $\{x(k)\}$ be defined by

$$x(k+1) = f(x(k), k) \qquad \text{for } k \geq t_0. \tag{1.4}$$

We say that $\{x(k)\}$ is a *dynamics* started with the initial condition $(t_0, x(t_0))$, or alternatively we say that $\{x(k)\}$ is the trajectory of the dynamics (1.4) started at *starting time* $t_0 \geq 0$ and *starting point* $x(t_0) \in X$. We refer to k as the time variable. Throughout the thesis, all the time variables are assumed to be non-negative integers.

For the dynamics (1.4), we say that a point $x \in X$ is an equilibrium point if $x = f(x, k)$ for any $k \geq 0$.

We say that the dynamics (1.4) is asymptotically stable if $\lim_{k \to \infty} x(k)$ exists for any initial condition $(t_0, x(t_0)) \in \mathbb{Z}^+ \times X$.

Suppose that X is a topological space. Then, we say that a function $V : X \to \mathbb{R}^+$ is a *Lyapunov function* for the dynamics (1.4) if $V(x)$ is a continuous function and also $V(x(k+1)) \leq V(x(k))$ for any initial condition $(t_0, x(t_0)) \in \mathbb{Z}^+ \times X$ and any $k \geq t_0$.

We say that a function $V : X \times \mathbb{Z}^+ \to \mathbb{R}^+$ is a *comparison function*[1] for the dynamics (1.4) if $V(x, k)$ is a continuous function of x for any $k \geq 0$ and also $V(x(k+1), k+1) \leq V(x(k), k)$ for any initial condition $(t_0, x(t_0)) \in \mathbb{Z}^+ \times X$ and any $k \geq t_0$.

As it can be seen from the provided definitions, the only difference between a Lyapunov function and a comparison function is that a Lyapunov function is a *time-invariant* function, whereas a comparison function could be a *time-varying* function.

References

1. J. Tsitsiklis, Problems in decentralized decision making and computation, Dissertation, Department of Electrical Engineering and Computer Science, MIT, 1984
2. J. Tsitsiklis, D. Bertsekas, M. Athans, Distributed asynchronous deterministic and stochastic gradient optimization algorithms. IEEE Trans. Autom. Control 31(9), 803–812 (1986)
3. A. Nedić, A. Ozdaglar, On the rate of convergence of distributed subgradient methods for multi-agent optimization, in *Proceedings of IEEE CDC*, 2007, pp. 4711–4716
4. I. Lobel, A. Ozdaglar, Distributed subgradient methods over random networks. IEEE Trans. Autom. Control **56**(6), 1291–1306 (2011)
5. S.S. Ram, A. Nedić, V.V. Veeravalli, Stochastic incremental gradient descent for estimation in sensor networks, in *Proceedings of Asilomar*, 2007
6. B. Touri, A. Nedić, S.S. Ram, Asynchronous stochastic convex optimization over random networks: Error bounds, *Information Theory and Applications Workshop (ITA)*, 2010
7. A. Nedić, A. Ozdaglar, P. Parrilo, Constrained consensus and optimization in multi-agent networks. IEEE Trans. Autom. Control **55**, 922–938 (2010)
8. S. Oh, S. Sastry, L. Schenato, A hierarchical multiple-target tracking algorithm for sensor networks, in *Proceedings of the 2005 IEEE International Conference on Robotics and Automation*, 2005, pp. 2197–2202
9. L. Schenato, G. Gamba., A distributed consensus protocol for clock synchronization in wireless sensor network," *46th IEEE Conference on Decision and Control*, 2007

---

[1] It is often required that Lyapunov functions and comparison functions be positive for non-equilibrium points. However, throughout this thesis we use Lyapunov functions and comparison functions in a loose sense of non-negative functions.

10. L. Schenato, F. Fiorentin, Average timesynch: a consensus-based protocol for time synchronization in wireless sensor networks, available at: http://paduaresearch.cab.unipd.it/2090/1/Necsys09_v2.pdf

11. P. Sommer, R. Wattenhofer, Gradient clock synchronization in wireless sensor networks, in *Proceedings of the 2009 International Conference on Information Processing in Sensor Networks*, 2009, pp. 37–48

12. U. Krause, Soziale dynamiken mit vielen interakteuren. eine problemskizze. *In Modellierung und Simulation von Dynamiken mit vielen interagierenden Akteuren*, 1997, pp. 37–51

13. R. Hegselmann, U. Krause, Opinion dynamics and bounded confidence models, analysis, and simulation. Journal of Artif Societies Social Simul vol. 5, 2002

14. F. Bullo, J. Cortés, S. Martínez, Distributed control of robotic networks, ser. Applied Mathematics Series. (Princeton University Press, 2009), electronically available at http://coordinationbook.info

15. J. Hajnal, The ergodic properties of non-homogeneous finite markov chains. Proc. Cambridge Philos. Soc. **52**(1), 67–77 (1956)

16. J. Wolfowitz, Products of indecomposable, aperiodic, stochastic matrices. Proc. Am. Math. Soc. **14**(4), 733–737 (1963)

17. J. Shen, *A geometric approach to ergodic non-homogeneous Markov chains*, ser. Lecture Notes in Pure and Applied Math vol. 212. (Marcel Dekker Inc., New York, 2000)

18. A. Kolmogoroff, Zur Theorie der Markoffschen Ketten. Mathematische Annalen **112**(1), 155–160 (1936)

19. M.H. DeGroot, Reaching a consensus. J. Am. Stat. Assoc. **69**(345), 118–121 (1974)

20. S. Chatterjee, E. Seneta, Towards consensus: Some convergence theorems on repeated averaging. J. Appl. Probab. **14**(1), 89–97 (1977)

21. A. Jadbabaie, J. Lin, S. Morse, Coordination of groups of mobile autonomous agents using nearest neighbor rules. IEEE Trans. Autom. Control **48**(6), 988–1001 (2003)

22. A. Olshevsky, J. Tsitsiklis, On the nonexistence of quadratic lyapunov functions for consensus algorithms. IEEE Trans. Autom. Control **53**(11), 2642–2645 (2008)

23. L. Moreau, Stability of multi-agent systems with time-dependent communication links. IEEE Trans. Autom. Control **50**(2), 169–182 (2005)

24. A. Olshevsky, J. Tsitsiklis, Convergence speed in distributed consensus and averaging. SIAM J. Control Optim. **48**(1), 33–55 (2008)

25. A. Olshevsky, Efficient information aggregation strategies for distributed control and signal processing, Dissertation, Massachussetts Institute of Technology, 2010

26. J. Rosenblatt, Products of independent identically distributed stochastic matrices. J. Math. Anal. Appl. **11**(1), 1–10 (1965)

27. K. Nawrotzki, Discrete open systems on markov chains in a random environment. I. Elektronische Informationsverarbeitung und Kybernetik **17**, 569–599 (1981)

28. K. Nawrotzki, Discrete open systems on markov chains in a random environment. II. Elektronische Informationsverarbeitung und Kybernetik **18**, 83–98 (1982)

29. R. Cogburn, On products of random stochastic matrices, In Random matrices and their applications, 1986, pp. 199–213

30. A. Tahbaz-Salehi, A. Jadbabaie, A necessary and sufficient condition for consensus over random networks. IEEE Trans. Autom. Control **53**(3), 791–795 (2008)

31. F. Fagnani, S. Zampieri, Randomized consensus algorithms over large scale networks. IEEE J. Sel. Areas Commun. **26**(4), 634–649 (2008)

32. A. Tahbaz-Salehi, A. Jadbabaie, Consensus over ergodic stationary graph processes. IEEE Trans. Autom. Control **55**(1), 225–230 (2010)

33. J. Lorenz, A stabilization theorem for continuous opinion dynamics. Phys A: Stat. Mech. Appl. **355**, 217–223 (2005)

34. J.M. Hendrickx, Graphs and networks for the analysis of autonomous agent systems, Dissertation, Université Catholique de Louvain, 2008

35. Y. Hatano, A. Das, M. Mesbahi, Agreement in presence of noise: Pseudogradients on random geometric networks, in *Proceedings of the 44th IEEE Conference on Decision and Control, and European Control Conference*, 2005, pp. 6382–6387
36. S. Patterson, B. Bamieh, A.E. Abbadi, Distributed average consensus with stochastic communication failures. IEEE Trans. Signal process. **57**, 2748–2761 (2009)
37. S. Boyd, A. Ghosh, B. Prabhakar, D. Shah, Randomized gossip algorithms. IEEE Trans. Inf. Theory **52**(6), 2508–2530 (2006)
38. A. Dimakis, A. Sarwate, M. Wainwright, Geographic gossip: Efficient averaging for sensor networks. IEEE Trans. Signal process. **56**(3), 1205–1216 (2008)
39. T. Aysal, M. Yildriz, A. Sarwate, A. Scaglione, Broadcast gossip algorithms for consensus. IEEE Trans. Signal process. **57**, 2748–2761 (2009)
40. R. Carli, F. Fagnani, P. Frasca, S. Zampieri, Gossip consensus algorithms via quantized communication. Automatica **46**(1), 70–80 (2010)
41. M. Huang, J. Manton, Stochastic approximation for consensus seeking: Mean square and almost sure convergence, in *Proceedings of the 46th IEEE Conference on Decision and Control*, 2007, pp. 306–311
42. M. Huang, J. Manton, Coordination and consensus of networked agents with noisy measurements: Stochastic algorithms and asymptotic behavior. SIAM J Control Optim. **48**(1), 134–161 (2009)
43. S. Kar, J. Moura, Distributed consensus algorithms in sensor networks: Link failures and channel noise. IEEE Trans. Signal Process. **57**(1), 355–369 (2009)
44. B. Touri, A. Nedić, Distributed consensus over network with noisy links, in *Proceedings of the 12th International Conference on Information Fusion*, 2009, pp. 146–154
45. J. Tsitsiklis, M. Athans, Convergence and asymptotic agreement in distributed decision problems. IEEE Trans. Autom. Control **29**(1), 42–50 (1984)
46. D. Bertsekas, J. Tsitsiklis, *Parallel and Distributed Computation: Numerical Methods* (Prentice-Hall Inc., Englewood Cliffs, 1989)
47. S. Li, T. Basar, Asymptotic agreement and convergence of asynchronous stochastic algorithms. IEEE Trans. Autom. Control **32**(7), 612–618 (1987)
48. R. Olfati-Saber, R. Murray, Consensus problems in networks of agents with switching topology and time-delays. IEEE Trans. Autom. Control **49**(9), 1520–1533 (2004)
49. W. Ren, R. Beard, Consensus seeking in multi-agent systems under dynamically changing interaction topologies. IEEE Trans. Autom. Control **50**(5), 655–661 (2005)
50. A. Kashyap, T. Basar, R. Srikant, Quantized consensus. Automatica **43**(7), 1192–1203 (2007)
51. A. Nedić, A. Olshevsky, A. Ozdaglar, J. Tsitsiklis, On distributed averaging algorithms and quantization effects. IEEE Trans. Autom. Control **54**(11), 2506–2517 (2009)
52. R. Carli, F. Fagnani, A. Speranzon, S. Zampieri, Communication constraints in the average consensus problem. Automatica **44**(3), 671–684 (2008)
53. R. Carli, F. Fagnani, P. Frasca, T. Taylor, S. Zampieri, Average consensus on networks with transmission noise or quantization, in *Proceedings of IEEE American Control Conference*, 2007, pp. 4189–4194
54. P. Bliman, A. Nedić, A. Ozdaglar, Rate of convergence for consensus with delays, in *Proceedings of the 47th IEEE Conference on Decision and Control*, 2008
55. B. Touri, A. Nedić, On ergodicity, infinite flow and consensus in random models. IEEE Trans. Autom. Control **56**(7), 1593–1605 (2011)
56. B. Touri, A. Nedić, When infinite flow is sufficient for ergodicity, in *Proceedings of 49th IEEE Conference on Decision and Control*, 2010, pp. 7479–7486
57. B. Touri, A. Nedić, Approximation and limiting behavior of random models, in *Proceedings of 49th IEEE Conference on Decision and Control*, 2010, pp. 2656–2663
58. B. Touri, A. Nedić, On existence of a quadratic comparison function for random weighted averaging dynamics and its implications, to appear in IEEE Conference on Decision and Control, 2011

59. B. Touri, A. Nedić, Alternative characterization of ergodicity for doubly stochastic chains, to appear in IEEE Conference on Decision and Control, 2011
60. B. Touri, A. Nedić, On approximations and ergodicity classes in random chains, 2010, under review
61. B. Touri, A. Nedić, On backward product of stochastic matrices, 2011, under review
62. B. Touri, T. Başar, A. Nedić, Averaging dynamics on general state spaces, 2011, technical report
63. B. Touri and A. Nedić, On product of random stochastic matrices, 2011, technical report

Chapter 2
Products of Stochastic Matrices and Averaging Dynamics

In this chapter, we introduce and review some of the results on products of stochastic matrices and averaging dynamics. Throughout this thesis, by a product of stochastic matrices, we mean *left product* of stochastic matrices. More precisely, let $\{A(k)\}$ be a stochastic chain. By left product of stochastic matrices, we mean the product of the form $A(k) \cdots A(t_0)$ where $k \geq t_0 \geq 0$ and k often approaches to infinity.

Generally, there are two alternative viewpoints for product of stochastic matrices. The first viewpoint is directly involved with the product itself. The second viewpoint is the dynamic system viewpoint, which is based on the study of the dynamics driven by such a product. In this section, we first discuss the two viewpoints and we show their equivalency. Then, we will present some results that are used in the thesis.

2.1 Averaging Dynamics: Two Viewpoints

Here, we discuss two viewpoints for averaging dynamics and show their equivalency.

2.1.1 Left Product of Stochastic Matrices

The first viewpoint builds on the convergence properties of the left product of stochastic matrices. Suppose that we have a sequence of stochastic matrices $\{A(k)\}$. Let

$$A(k : s) = A(k - 1) \cdots A(s) \qquad \text{for } k > s \geq 0,$$

and $A(s : s) = I$. Let us define the concepts of weak and strong ergodicity as appeared in [1].

B. Touri, *Product of Random Stochastic Matrices and Distributed Averaging*,
Springer Theses, DOI: 10.1007/978-3-642-28003-0_2,

Definition 2.1 (*Ergodicity* [1]) Let $\{A(k)\}$ be a chain of stochastic matrices. We say that $\{A(k)\}$ is weakly ergodic if $\lim_{k\to\infty}(A_{i\ell}(k:t_0) - A_{j\ell}(k:t_0)) = 0$ for any $t_0 \geq 0$ and $i, j, \ell \in [m]$.

We say that $\{A(k)\}$ is strongly ergodic, if $\lim_{k\to\infty} A(k:t_0) = ev^T(t_0)$ *for any* $t_0 \geq 0$, where $v(t_0)$ is a stochastic vector in $\mathbb{R}^m$.

In words, we say that $\{A(k)\}$ is weakly ergodic if for any *starting time* $t_0 \geq 0$ the difference between any two rows of the product $A(k:t_0)$ goes to zero as k approaches infinity. Likewise, we say that $\{A(k)\}$ is strongly ergodic if for any *starting time* $t_0 \geq 0$, the product $A(k:t_0)$ approaches a stochastic matrix with identical rows, i.e. a rank one stochastic matrix.

Note that in the definition of strong ergodicity the requirement that $v(t_0)$ should be stochastic is redundant. This follows directly from the fact that the ensemble of $m \times m$ stochastic matrices is a closed set in $\mathbb{R}^{m\times m}$, which can be verified by noting that

$$\mathbb{S}_m = \{A \in \mathbb{R}^{m\times m} \mid A \geq 0, Ae = e\}. \tag{2.1}$$

Thus, $\mathbb{S}_m$ is the intersection of two closed sets $\{\mathcal{A} \in \mathbb{R}^{m\times m} \mid A \geq 0\}$ and $\{A \in \mathbb{R}^{m\times m} \mid Ae = e\}$, implying that $\mathbb{S}_m$ is a closed subset of $\mathbb{R}^{m\times m}$. Hence, $\lim_{k\to\infty} A(k:t_0) = ev^T(t_0)$ automatically implies that $v(t_0)$ is a stochastic vector.

Example 2.1 As a simple example for ergodicity, let $\{A(k)\}$ be a chain of stochastic matrices such that $A(k)$ is equal to the rank one stochastic matrix $\frac{1}{m}J$ for infinitely many indices $\cdots > k_t > \cdots > k_2 > k_1$. Note that for any stochastic matrix A, we have $A[\frac{1}{m}J] = \frac{1}{m}J$. Also, we have $[\frac{1}{m}J]A = \frac{1}{m}ee^T A = e[\frac{1}{m}e^T A]$. Note that $[\frac{1}{m}e^T A]$ is a stochastic vector, since $\frac{1}{m}e^T Ae = \frac{1}{m}e^T e$. Thus, for any $t_0 \geq 0$, if t is large enough such that $k_t > t_0$, then we have

$$A(k:t_0) = A(k:k_t)A(k_t:t_0) = \frac{1}{m}JA(k_t:t_0) = e[\frac{1}{m}e^T A(k_t:t_0)],$$

for all $k > k_t$. This implies that $\lim_{k\to\infty} A(k:t_0) = ev^T(t_0)$ for any $t_0 \geq 0$, where $v(t_0) = \frac{1}{m}A^T(k_t:t_0)e$. Thus such a chain is strongly ergodic.

Note that if $\lim_{k\to\infty} A(k:t_0) = ev^T(t_0)$, then we have $\lim_{k\to\infty}(A_{ij}(k:t_0) - A_{\ell j}(k:t_0)) = v_j(t_0) - v_j(t_0) = 0$. Thus, strong ergodicity implies weak ergodicity. In [1], it is proven that the reverse implication also holds.

Theorem 2.1 ([1], *Theorem* 1) *Weak ergodicity is equivalent to strong ergodicity.*

Since weak and strong ergodicity are the same concepts, we simply refer to this property as *ergodicity*. Thus, there are two equivalent viewpoints to ergodicity: the first viewpoint is that the product $A(k) \cdots A(t_0)$ converges to a matrix with identical stochastic rows, or alternatively, the difference between any two rows of the product $A(k) \cdots A(t_0)$ converges to zero as k goes infinity, for any choice of starting time t_0. This is the viewpoint to ergodicity and averaging based on the left product of stochastic matrices.

Another related concept that is based on the limiting behavior of the product of stochastic matrices is *consensus* as defined below.

Definition 2.2 (*Consensus*) We say that a chain $\{A(k)\}$ admits consensus if $\lim_{k\to\infty} A(k:0) = ev^T$ for some stochastic vector $v \in \mathbb{R}^m$.

The reason that such property is called *consensus* will be clear once we discuss the averaging dynamics from the dynamic system viewpoint.

Note that a chain $\{A(k)\}$ is ergodic if and only if $\{A(k)\}_{k\geq t_0}$ admits consensus for any $t_0 \geq 0$. Thus ergodicity highly depends on the future (tale) of the chain $\{A(k)\}$. However, this may not be the case for consensus as seen from the following example.

Example 2.2 Consider the chain $\{A(k)\}$ defined by $A(0) = \frac{1}{m}J$ and $A(k) = I$ for $k \geq 1$. Then, for any $k > 0$, we have $A(k:0) = \frac{1}{m}J$ implying that $\{A(k)\}$ admits consensus. However, note that for any starting time $t_0 \geq 1$, and any $k > t_0$, we have $A(k:t_0) = I$ implying that $\{A(k)\}$ is not ergodic.

Although consensus may not be dependent on the future and, in general, it does not imply ergodicity but under a certain condition the reverse holds.

Lemma 2.1 *Let $\{A(k)\}$ be a chain of stochastic matrices and suppose that $A(k)$ is invertible for any $k \geq 0$. Then, $\{A(k)\}$ admits consensus if and only if $\{A(k)\}$ is ergodic.*

Proof Note that if $A(k)$ is invertible for any $k \geq 0$, then for any $t_0 \geq 0$, the limit $\lim_{k\to\infty} A(k:t_0)$ exists if and only if $\lim_{k\to\infty} A(k:0)$ exists. This follows from the fact that for $k > t_0$, we have $A(k:t_0) = A(k:0)A^{-1}(t_0:0)$.

Now, if $\{A(k)\}$ admits consensus, then $\lim_{k\to\infty} A(k:0)$ is a rank one matrix. Therefore, for any $t_0 \geq 0$, the matrix $[\lim_{k\to\infty} A(k:t_0)]A(t_0:0)$ is a rank one matrix and since $A(t_0:0)$ is a full-rank matrix, it implies that $\lim_{k\to\infty} A(k:t_0)$ is a rank one matrix. But this holds for any $t_0 \geq 0$, implying that $\{A(k)\}$ is ergodic. The reverse implication follows by the definition of ergodicity and consensus.   **Q.E.D**

2.1.2 Dynamic System Viewpoint

Here, we discuss the dynamic system viewpoint to the averaging dynamics and we show that it is equivalent to the previously discussed viewpoint.

Let $\{A(k)\}$ be a chain of stochastic matrices. Let $t_0 \geq 0$ and let $x(t_0) \in \mathbb{R}^m$ be arbitrary. Let us define:

$$x(k+1) = A(k)x(k), \qquad \text{for } k \geq t_0. \tag{2.2}$$

We say that $\{x(k)\}$ is a dynamics driven by $\{A(k)\}$. We also say that $t_0 \geq 0$ is the *starting time* and $x(t_0) \in \mathbb{R}^m$ is the *starting point* of the dynamics. We refer to $(t_0, x(t_0)) \in \mathbb{Z}^+ \times \mathbb{R}^m$ as an *initial condition* for the dynamics.

A side remark about dynamics (2.2) is that the chain $\{A(k)\}$ may depend on the history of $\{x(k)\}$, i.e. $A(k)$ may be some function of the time k and the history of the dynamics $x(t_0), \ldots, x(k)$. In this case, we extract the process $\{B(k)\}$ by letting $B(k) = A(k, x(t_0), \ldots, x(k))$ and study the dynamics (2.2) for the chain $\{B(k)\}$ with arbitrary initial condition.

Also, note that any point in the set

$$\mathbf{C} = \{\lambda e \mid \lambda \in \mathbb{R}\},$$

which is the line passing through the all-one vector e, is an equilibrium point for the dynamics (2.2). We refer to this line as the *consensus subspace*.

Now, consider a starting time $t_0 \geq 0$ and a starting point $x(t_0) \in \mathbb{R}^m$, and consider the corresponding dynamics $\{x(k)\}$ driven by a stochastic chain $\{A(k)\}$. At each time $k \geq t_0$, we have $x_i(k+1) = \sum_{j=1}^{m} A_{ij}(k)x_j(k)$. But $A_i(k)$ is a stochastic vector and hence, $x_i(k+1)$ is simply a *weighted average* of the scalars $\{x_1(k), \ldots, x_m(k)\}$. Thus, the m coordinates of the vector $x(k+1)$ are nothing but m weighted averages of the coordinates of $x(k)$. This motivates the name *weighted averaging dynamics* for the dynamics. Based on this observation, an intuitive way of describing dynamics (2.2) is to consider the set $[m]$ as a set of m *agents* and let $x_i(t_0) \in \mathbb{R}$ to be a scalar representing the initial opinion of the ith agent about an issue. Then, at each time $k \geq t_0$, agents share their opinions and agent i's opinion will evolve by averaging the observed opinions at time k. Although such a model of opinion dynamics among a set of agents is hypothetical, we refer to this interpretation of the dynamics (2.2) as the *opinion dynamics* viewpoint to dynamics (2.2).

For any dynamics $\{x(k)\}$ and any $k \geq t_0$, we have $x(k) = A(k : t_0)x(t_0)$. Therefore, if $\{A(k)\}$ is an ergodic chain (as defined in Definition 2.1), for any initial condition $(t_0, x(t_0)) \in \mathbb{Z}^+ \times \mathbb{R}^m$, we have:

$$\lim_{k \to \infty} x(k) = \lim_{k \to \infty} A(k : t_0)x(t_0) = ev^T(t_0)x(t_0) = c(t_0)e,$$

where $c(t_0) = v^T(t_0)x(t_0)$ is a constant depending on the initial condition. Thus, if $\{A(k)\}$ is an ergodic chain, then for any initial condition $(t_0, x(t_0)) \in \mathbb{Z}^+ \times \mathbb{R}^m$, any dynamics $\{x(k)\}$ will converge to some point in the consensus subspace $\mathbf{C}$. This implies that $\lim_{k \to \infty}(x_i(k) - x_j(k)) = 0$ for any $i, j \in [m]$ and any initial condition $(t_0, x(t_0)) \in \mathbb{Z}^+ \times \mathbb{R}^m$. In fact, the reverse implication also holds.

Theorem 2.2 *A stochastic chain $\{A(k)\}$ is ergodic if and only if* $\lim_{k \to \infty} (x_i(k) - x_j(k)) = 0$ *for every initial condition $(t_0, x(t_0)) \in \mathbb{Z}^+ \times \mathbb{R}^m$ and all $i, j \in [m]$.*

Furthermore, for ergodicity, it suffices that $\lim_{k \to \infty} (x_i(k) - x_j(k)) = 0$ *for any starting time $t_0 \geq 0$ and $x(t_0) = e_\ell$ for all $\ell \in [m]$.*

Proof The fact that the ergodicity of $\{A(k)\}$ implies $\lim_{k \to \infty}(x_i(k) - x_j(k)) = 0$ for any initial condition $(t_0, x(t_0)) \in \mathbb{Z}^+ \times \mathbb{R}^m$ and all $i, j \in [m]$ follows from the above discussion. For the converse implication, suppose that $\lim_{k \to \infty}(x_i(k) - x_j(k)) = 0$ for all, $i, j \in [m]$, and every starting time $t_0 \geq 0$ and any starting point $x(t_0) = e_\ell$

where $\ell \in [m]$. For such a starting time, we have $x(k) = A(k : t_0)e_\ell = A^\ell(k : t_0)$ and hence, $x_i(k) = A_{i\ell}(k : t_0)$ and $x_j(k) = A_{j\ell}(k : t_0)$. Therefore, $\lim_{k\to\infty}(x_i(k) - x_j(k)) = 0$ for $x(t_0) = e_\ell$ if and only if $\lim_{k\to\infty}(A_{i\ell}(k : t_0) - A_{j\ell}(k : t_0)) = 0$. Thus, if $\lim_{k\to\infty}(x_i(k) - x_j(k)) = 0$ for any starting time $t_0 \geq 0$ and any starting point of the form $x(t_0) = e_\ell$ for $\ell \in [m]$, then $\{A(k)\}$ is an ergodic chain.   **Q.E.D**

Using a similar argument, the following result follows immediately.

Theorem 2.3 *A stochastic chain $\{A(k)\}$ admits consensus if and only if*

$$\lim_{k\to\infty} \big(x_i(k) - x_j(k)\big) = 0$$

for the starting time $t_0 = 0$ and every starting point $x(0) \in \mathbb{R}^m$.
Furthermore, to show that $\{A(k)\}$ admits consensus, it suffices that

$$\lim_{k\to\infty} \big(x_i(k) - x_j(k)\big) = 0$$

for all starting points $x(0) = e_\ell$ with $\ell \in [m]$.

This result explains why the property defined in Definition 2.2 is referred to as consensus. Basically, admitting consensus means that for any initial opinion of the set of agents at time 0, the dynamics (2.2) will lead to consensus, i.e. the difference $x_i(k) - x_j(k)$ between the opinions of any two agents $i, j \in [m]$ goes to zero as k goes to infinity.

2.1.3 Uniformly Bounded and B-Connected Chains

For the study of the averaging dynamics, it is often assumed that the positive entries of a stochastic chain $\{A(k)\}$ are bounded below by some positive scalar $\gamma > 0$ which makes the study of those chains more convenient. We refer to this property as uniform boundedness property.

Definition 2.3 We say that a stochastic chain $\{A(k)\}$ is uniformly bounded if there exist a scalar $\gamma > 0$ for which $A_{ij}(k) \geq \gamma$ for all $i, j \in [m]$ and $k \geq 0$ such that $A_{ij}(k) > 0$.

Some consensus and ergodicity results for deterministic weighted averaging dynamics rely on the existence of a periodical connectivity of the graphs associated with the matrices. We refer to this property as B-connectedness property.

Definition 2.4 For a stochastic chain $\{A(k)\}$, let $G(k) = ([m], \mathcal{E}(k))$ be the graph induced by the positive entries of $A(k)$ for any $k \geq 0$. For $B \geq 1$, we say $\{A(k)\}$ is a *B-connected chain* if

(a) $\{A(k)\}$ is uniformly bounded,
(b) for any time $k \geq 0$ and $i \in [m]$, we have $A_{ii}(k) > 0$, and

(c) the graph:

$$([m], \mathcal{E}(Bk) \cup \mathcal{E}(Bk + 1) \cup \cdots \cup \mathcal{E}(B(k + 1) - 1)),$$

is strongly connected for *any* $k \geq 0$.

The following result shows that a B-connected chain is ergodic. Furthermore, using the result, we can provide some bounds on the limiting matrix $\lim_{k \to \infty} A(k : t_0)$.

Theorem 2.4 ([2], *Lemma 5.2.1*) *Let* $\{A(k)\}$ *be a B-connected chain. Then, the following results hold*:

(a) $\{A(k)\}$ *is ergodic, i.e. for any starting time* $t_0 \geq 0$, *we have* $\lim_{k \to \infty} A(k : t_0) = ev^T(t_0)$ *for a stochastic vector* $v(t_0) \in \mathbb{R}^m$.
(b) *There is a constant* $\eta \geq \gamma^{(m-1)B}$, *which is independent of* t_0, *such that* $v(t_0) \geq \eta$ *for all* $t_0 \geq 0$.
(c) *We have*

$$\max_{i,j} |A_{ij}(k : t_0) - [ev^T(t_0)]_{ij}| \leq q\mu^{k-t_0},$$

where $\mu \in (0, 1)$ *and* $q \in \mathbb{R}^+$ *are some constants not depending on* t_0.

2.1.4 Birkhoff-von Neumann Theorem

Consider the set of doubly stochastic matrices

$$\mathcal{D} = \{A \in \mathbb{R}^{m \times m} \mid A \geq 0, \ Ae = A^T e = e\}. \tag{2.3}$$

The given description in Eq. (2.3) clearly shows that the set of doubly stochastic matrices is a polyhedral set in $\mathbb{R}^{m \times m}$. On the other hand, a permutation matrix P is an extreme point of this set, i.e. we cannot write $P = \epsilon A + (1 - \epsilon)B$ for some distinct $A, B \in \mathcal{D}$ and $\epsilon \in (0, 1)$.

The Birkhoff-von Neumann theorem asserts that, in fact, permutation matrices are the only extreme points of $\mathcal{D}$.

Theorem 2.5 (Birkhoff-von Neumann Theorem [3], p. 527) *Let A be a doubly stochastic matrix. Then, A can be written as a convex combination of the permutation matrices, i.e. there exists scalars* $q_1, \ldots, q_{m!} \in \mathbb{R}^+$ *such that*

$$A = \sum_{\xi=1}^{m!} q_\xi P^{(\xi)},$$

where $\sum_{\xi \in [m!]} q_\xi = 1$.

We use this result in Chap. 6 to provide an alternative characterization of ergodicity for doubly stochastic chains.

References

1. S. Chatterjee, E. Seneta, Towards consensus: Some convergence theorems on re–peated averaging. J. Appl. Probab. **14**(1), 89–97 (1977)
2. J. Tsitsiklis, Problems in decentralized decision making and computation, Ph.D. dissertation, Department of Electrical Engineering and Computer Science, MIT, 1984
3. R.A. Horn, C.R. Johnson, *Matrix Analysis* (Cambridge University Press, Cambridge, 1985)

Chapter 3
Ergodicity of Random Chains

In this chapter, we build the framework for the study of random averaging dynamics. We also introduce some of the central objects of this work such as infinite flow property and infinite flow graph.

The structure of this chapter is as follows: In Sect. 3.1, we discuss the random framework for study of product of random stochastic matrices or equivalently, random averaging dynamics. In Sect. 3.2, we discuss the infinite flow property and we prove the necessity of the infinite flow for ergodicity of any chain. In Sect. 3.3, we relate the infinite flow property to the connectivity of a graph, the infinite flow graph associated with the given stochastic chain. We also introduce an ℓ_1-approximation of a stochastic chain and we prove that such an approximation preserves the ergodic behavior of a stochastic chain. Using this result, we provide an extension of the necessity of the infinite flow property to non-ergodic chains.

3.1 Random Weighted Averaging Dynamics

We study the product of stochastic matrices, or averaging dynamics (2.2) in a general setting of random dynamics. To do this, let $(\Omega, \mathcal{F}, \mathrm{pr}\,(\cdot))$ be a probability space and let $\{\mathcal{F}_k\}$ be a filtration on $(\Omega, \mathcal{F})$. In what follows, we provide the definition of a random chain and an adapted random chain which are among the central objects of this thesis.

Definition 3.1 We say that $\{W(k)\}$ is a random stochastic chain, or simply a random chain, if

(a) $W(k)$ is a random matrix process, i.e., $W(k)$ is a random matrix for any $k \geq 0$, and
(b) $W(k)$ is a stochastic matrix almost surely for any $k \geq 0$.

Furthermore, if $W(k)$ is measurable with respect to $\mathcal{F}_{k+1}$, i.e. $W_{ij}(k)$ is measurable with respect to $\mathcal{F}_{k+1}$ for all $i, j \in [m]$ and any $k \geq 0$, with an abuse of notation, we

B. Touri, *Product of Random Stochastic Matrices and Distributed Averaging*,
Springer Theses, DOI: 10.1007/978-3-642-28003-0_3,
© Springer-Verlag Berlin Heidelberg 2012

say that $\{W(k)\}$ is a random chain adapted to $\{\mathcal{F}_k\}$, or simply $\{W(k)\}$ is an adapted random chain.

If the random matrices in $\{W(k)\}$ are independently distributed, we say that $\{W(k)\}$ is an independent chain. If furthermore, the random matrices have an identical distribution we say that $\{W(k)\}$ is an identically independently distributed chain, and we use the abbreviation i.i.d. to denote such a property.

For a given random chain $\{W(k)\}$, we consider the random dynamics of the following form:

$$x(k+1) = W(k)x(k), \tag{3.1}$$

started at a starting time $t_0 \geq 0$ and a starting point $x(t_0, \omega) = v \in \mathbb{R}^m$ for all $\omega \in \Omega$. Note that, in this case, we have $x(t_0, \omega) = v$ for any $\omega \in \Omega$ and since $\emptyset$ and Ω belong to any σ-algebra, it follows that $x(t_0)$ is measurable with respect to $\mathcal{F}_{t_0}$. Since for any adapted chain $\{W(k)\}$, $W(k)$ is measurable with respect to $\mathcal{F}_{k+1}$, it follows that any random dynamics $\{x(k)\}$ is adapted to the filtration $\{\mathcal{F}_k\}$. With an abuse of notation, instead of referring to the starting point $x(t_0)$ as a measurable function $x(t_0, \omega) = v \in \mathbb{R}^m$ for all $\omega \in \Omega$, we often simply say that the random dynamics $\{x(k)\}$ is started at the point $x(t_0) \in \mathbb{R}^m$.

We refer to $\{x(k)\}$ as a *random dynamics* driven by $\{W(k)\}$ and, as in the case of the deterministic dynamics, we refer to $(t_0, x(t_0))$ as an initial condition for the dynamics $\{x(k)\}$.

To avoid confusion between a deterministic chain and a random chain, we use the first alphabet letters to denote deterministic chains (such as $\{A(k)\}$ and $\{B(k)\}$) and the nearly last alphabet letters to represent random chains (such as $\{W(k)\}$ and $\{U(k)\}$).

For our further development, we let

$$\bar{W}(k) = \mathsf{E}[W(k) \mid \mathcal{F}_k],$$

and refer to $\{\bar{W}(k)\}$ as the *expected chain of* $\{W(k)\}$.

Note that any deterministic chain $\{A(k)\}$ can be considered as an independent random chain and hence, the dynamics (2.2) is an instance of the random dynamics (3.1).

3.2 Ergodicity and Infinite Flow Property

In this section, we first discuss ergodicity and consensus for random averaging dynamics. Then, we introduce the concept of infinite flow property and investigate the relation between ergodicity and infinite flow property.

The concepts of ergodicity and consensus can be naturally generalized to random chains. For a random model $\{W(k)\}$, there are subsets of the underlying probability space on which ergodicity and consensus happen. Specifically, let $\mathscr{E}$ and $\mathscr{C}$ be the subsets of Ω on which the ergodicity and consensus happen, respectively. Then, $\mathscr{E}$ and $\mathscr{C}$ are measurable subsets.

Lemma 3.1 *Let $\{W(k)\}$ be a random chain on a probability space $(\Omega, \mathcal{F})$. Then, the subsets $\mathscr{E}$ and $\mathscr{C}$ over which ergodicity and consensus happen are measurable subsets.*

Proof As shown in Theorem 2.1, $\mathscr{E}$ is the subset of Ω over which

$$\lim_{k \to \infty} \left(x_i(k, \omega) - x_j(k, \omega) \right) = 0,$$

for all starting times $t_0 \geq 0$ and starting points $x(t_0, \omega) = e_\ell$ for all $\ell \in \mathbb{R}^m$. Thus,

$$\mathscr{E} = \bigcap_{t_0=0}^{\infty} \left(\bigcap_{\ell=1}^{m} \{\omega \mid \lim_{k \to \infty} \left(x_i(k, \omega) - x_j(k, \omega) \right) = 0 \quad \text{for all } i, j \in [m], x(t_0, \omega) = e_\ell \} \right).$$

But all $W_{ij}(k)$ s are Borel-measurable with respect to $\mathcal{F}$ and hence, by Lemma A.1, for fixed $i, j, \ell \in [m]$, the set $\{\omega \mid \lim_{k \to \infty}(x_i(k, \omega) - x_j(k, \omega)) = 0\}$ where $x(t_0) = e_\ell$ is measurable. Since $\mathscr{E}$ is an intersection of finitely many measurable sets, it follows that $\mathscr{E}$ is a measurable set itself.

Similarly, for the consensus event $\mathscr{C}$, we have

$$\mathscr{C} = \bigcap_{\ell=1}^{m} \{\omega \mid \lim_{k \to \infty} \left(x_i(k, \omega) - x_j(k, \omega) \right) = 0 \quad \text{for all } i, j \in [m], x(0, \omega) = e_\ell \},$$

and, using a similar argument, we conclude that $\mathscr{C} \in \mathcal{F}$. **Q.E.D**

We refer to the events $\mathscr{E}$ and $\mathscr{C}$ as ergodicity and consensus events, respectively, and we say that a random model is *ergodic* (*admits consensus*) if the event $\mathscr{E}$ ($\mathscr{C}$) happens almost surely.

Now, let us discuss the concept of the infinite flow property which is closely related to ergodicity. Consider a B-connected chain $\{A(k)\}$ as defined in Definition 2.4. Consider a subset $S \subset [m]$. By assumption (c) of Definition 2.4, it follows that there is at least one edge connecting a vertex in S to a vertex in $\bar{S}$ in the time interval $[kB, (k + 1)B)$ for any $k \geq 0$. On the other hand, by the uniform bounded-ness property of $\{A(k)\}$, it follows that $\sum_{t=kB}^{(k+1)B-1} A_S(t) \geq \gamma > 0$ for some $\gamma > 0$. Thus, for any $S \subset [m]$, the B-connectivity assumption on $\{A(k)\}$ implies that $\sum_{t=0}^{\infty} A_S(t) = \infty$. Note that this property happens for any $S \subset [m]$. Interestingly enough, any ergodic chain $\{A(k)\}$ exhibits the same behavior which will be shown subsequently. Before proving this result, let us identify this property as the *infinite flow property*.

Definition 3.2 We say that a stochastic chain $\{A(k)\}$ has the infinite flow property if $\sum_{k=0}^{\infty} A_S(k) = \infty$ for any non-trivial $S \subset [m]$.

To motivate the necessity of infinite flow property for ergodicity, consider the opinion dynamics interpretation of Eq.(1.4). One can interpret $A_{ij}(k)$ as the *credit* that agent i gives to agent j's opinion at time k. Therefore, $\sum_{i \in S, j \in \bar{S}} A_{ij}(k)$ can be interpreted as the credit that the subgroup $S \subset [m]$ of agents gives to the opinions of the agents that are outside of S (the agents in $\bar{S}$) at time k. Similarly, $\sum_{i \in \bar{S}, j \in S} A_{ij}(k)$ is the credit that the remaining agents, i.e. those agents in $\bar{S}$, give to the opinion of agents in S at time $k \geq 0$. The intuition behind the concept of infinite flow property is that without having infinite accumulated credit between groups S and $\bar{S}$ of agents, we cannot have an agreement among the agents in the sets S and $\bar{S}$ for any starting time t_0 of the opinion dynamic and for any initial opinion profile $x(t_0)$ of the agents. In other words, the infinite flow property is required to ensure necessary flow of credits between the agents in S and $\bar{S}$ for any non-trivial subset $S \subset [m]$.

One of the key-observations in our development is that the infinite flow property is a necessary condition for ergodicity. There are several ways of proving this result. Here, we provide a non-standard way of proving it using randomization technique. In Theorem 7.1, we provide an algebraic proof for this result on a general state space. Also, an extension of this result will be proven in Lemma 3.6. The proof that is presented here is based on the geometric structure of the set of $m \times m$ stochastic matrices $\mathbb{S}_m$. Note that

$$\mathbb{S}_m = \{A \in \mathbb{R}^{m \times m} \mid Ae = e, A \geq 0\}.$$

This shows that the set $\mathbb{S}_m$ is a polyhedral set. Let $\mathbf{M} = \{M^{(\xi)} \mid \xi \in [m^m]\}$ be the ensemble of matrices which has one entry equal to one at each row. It can be seen that each $M^{(\xi)}$ is an extreme point of $\mathbb{S}_m$. Furthermore, it can be proven that in fact these points are the only extreme points of $\mathbb{S}_m$ and, hence, we can write any stochastic matrix as a convex combination of the points in $\mathbf{M}$. Using this, we prove the necessity of the infinite flow property.

Theorem 3.1 *Infinite flow property is necessary for ergodicity of any deterministic chain.*

Proof Let $\{A(k)\}$ be a stochastic chain with $\sum_{k=0}^{\infty} A_S(k) < \infty$ for some non-trivial $S \subset [m]$. Then, for any $k \geq 0$, we can write $A(k)$ as a convex combination of the elements in $\mathbf{M}$, i.e., there exist non-negative scalars $p_1(k), \ldots, p_{m^m}(k)$ such that

$$A(k) = \sum_{\xi=1}^{m^m} p_\xi(k) M^{(\xi)},$$

and $\sum_{\xi=1}^{m^m} p_\xi(k) = 1$.

Now, let $\{W(k)\}$ be an independent random chain defined by:

$$W(k) = M^{(\xi)} \qquad \text{with probability } p_\xi(k).$$

This implies that $\mathsf{E}[W(k)] = \sum_{\xi=1}^{m^m} p_\xi M^{(\xi)} = A(k)$. Now, let

$$\mathbf{M}_S = \{M \in \mathbf{M} \mid M_S \geq 1\},$$

which is the subset of $\mathbf{M}$ containing matrices M with entries $M_{ij} = 1$ or $M_{ji} = 1$ for some $i \in S$ and $j \in \bar{S}$. The important feature of the set $\mathbf{M}_S$ is that for any $M \notin \mathbf{M}_S$, we have $M e_S = e_S$ where $e_S = \sum_{i \in S} e_i$. This is true because if $M \notin \mathbf{M}_S$, then $M_{ij} = M_{ji} = 0$ for all $i \in S$ and $j \notin \bar{S}$.

Now, note that $A_S(k) = \Pr(W(k) \in \mathbf{M}_S)$. Hence, if

$$\sum_{k=0}^{\infty} A_S(k) = \sum_{k=0}^{\infty} \Pr(W(k) \in \mathbf{M}_S) < \infty,$$

for some $S \subset [m]$, then, by the Second Borel-Cantelli lemma (Lemma A.6) it follows that $\Pr(W(k) \in \mathbf{M}_S \text{ i.o. }) = 0$. This means that for almost any sample path of $\{W(k)\}$, there exists a large enough random time T_S such that $T_S < \infty$ a.s. and $W(k) \in \mathbf{M} \setminus \mathbf{M}_S$ for $k \geq T_S$. Thus, we almost surely have $\emptyset = \bigcap_{t=0}^{\infty}(T_S \geq t)$. By continuity of measure, it follows that there exists a large enough $t_0 \geq 0$ such that $\Pr(T_S \geq t_0) \leq \frac{1}{3}$. Now, let $x(t_0) = e_S$ and let $\{x(k)\}$ be the random dynamics driven by $\{W(k)\}$ and let $\{\bar{x}(k)\}$ be the deterministic dynamics driven by $\{A(k)\}$ started at time t_0 at the starting point $\bar{x}(t_0) = e_S$.

Note that since $A(k) = \mathsf{E}[W(k)]$ and $\{W(k)\}$ is an independent random chain, it follows that $\bar{x}(k) = \mathsf{E}[W(k)]$. But for any $\omega \in \{T_S < t_0\}$, we have $W(k) \notin \mathbf{M}_S$ for $k \geq t_0$. Thus, for $\omega \in \{T_S < t_0\}$, the dynamics $\{x(k)\}$ is a static sequence $\{e_S\}$ which follows from the fact that $M e_S = e_S$ for $M \notin \mathbf{M}_S$. On the other hand, for any $k \geq t_0$, we have $x_i(k) \in [0, 1]$ almost surely and hence, for any $i \in S$, we have

$$\bar{x}_i(k) = \mathsf{E}\big[x_i(k)\mathbf{1}_{\{T_S < t_0\}}\big] + \mathsf{E}\big[x_i(k)\mathbf{1}_{\{T_S < t_0\}^c}\big] \geq \mathsf{E}\big[x_i(k)\mathbf{1}_{\{T_S < t_0\}}\big] \geq \frac{2}{3}.$$

Similarly, for $j \in \bar{S}$ we have $\bar{x}_j(k) \leq \frac{1}{3}$ which implies that for any $k \geq t_0$, we have $\bar{x}_i(k) - \bar{x}_j(k) \geq \frac{1}{3}$. Thus, we found some $(t_0, \bar{x}(t_0)) \in \mathbb{Z}^+ \times \mathbb{R}^m$ such that for the dynamics $\{\bar{x}(k)\}$ driven by $\{A(k)\}$, we have $\liminf_{k \to \infty} (\bar{x}_i(k) - \bar{x}_j(k)) > 0$ for some $i, j \in [m]$ which implies that $\{A(k)\}$ is not ergodic.   **Q.E.D**

Theorem 3.1 shows that the minimum requirement for a chain $\{A(k)\}$ to be an ergodic chain is having the infinite flow property. In our forthcoming chapter, we develop general conditions for which the reverse implication is also true.

Now, let us discuss the infinite flow property for a random chain $\{W(k)\}$. As in the cases of ergodicity and consensus events, for a random chain $\{W(k)\}$, the infinite flow property holds on a measurable subset of the probability space $(\Omega, \mathcal{F})$.

Lemma 3.2 *For a random chain $\{W(k)\}$ on a probability space $(\Omega, \mathcal{F})$, the set $\mathscr{F}$ on which the infinite flow property happens is a measurable set.*

Proof By the definition of the infinite flow property, we have

$$\mathscr{F} = \bigcap_{\substack{S \subset [m] \\ S \neq \emptyset}} \{\omega \mid \sum_{k=0}^{\infty} W_S(k) = \infty\}.$$

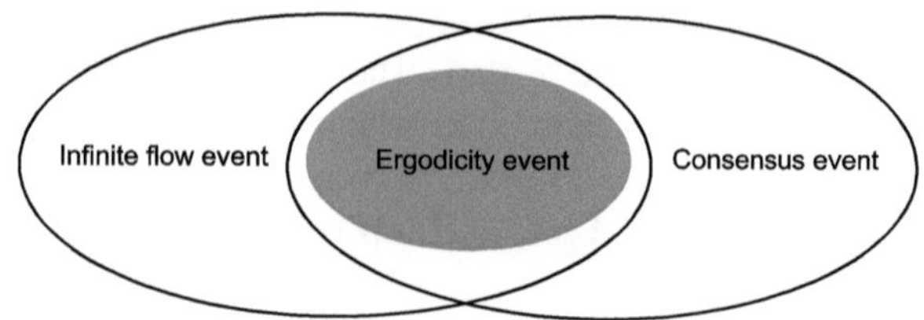

Fig. 3.1 Relation between ergodicity, consensus, and infinite flow events for a general random chain

Each $W_S(k)$ is a measurable function, and so is $\sum_{k=0}^{t} W_S(k)$. Therefore, by Lemma A.1, we conclude that $\mathscr{F}$ is a measurable set in $(\Omega, \mathcal{F})$.   **Q.E.D**

We refer to $\mathscr{F}$ as the infinite flow event, and we say that a random chain $\{W(k)\}$ has the infinite flow property if $\{W(k)\}$ has infinite flow property almost surely.

By Theorem 3.1, for a general random chain $\{W(k)\}$, we have $\mathscr{E} \subseteq \mathscr{F}$. Also, by the definition of ergodicity and consensus we have $\mathscr{E} \subseteq \mathscr{C}$. This situation is depicted in Fig. 3.1.

3.3 Infinite Flow Graph and ℓ_1-Approximation

So far, we showed that the infinite flow property is necessary for ergodicity of any stochastic chain. In this section, we show that an extension of this necessary condition holds for *non-ergodic* chains.

For our development, let us define the infinite flow graph associated with a stochastic chain $\{A(k)\}$.

Definition 3.3 (*Infinite Flow Graph*) For a deterministic chain $\{A(k)\}$ we define the infinite flow graph to be the graph $G^\infty = ([m], \mathcal{E}^\infty)$ with

$$\mathcal{E}^\infty = \{\{i, j\} \mid \sum_{k=0}^{\infty} \big(A_{ij}(k) + A_{ji}(k)\big) = \infty, i \neq j \in [m]\}.$$

In other words, we connect two distinct vertices $i, j \in [m]$ if the link between i and j carry infinite flow over the time in either of the two directions. The next result shows that the connectivity of the infinite flow graph is equivalent to the infinite flow property.

Lemma 3.3 *A chain $\{A(k)\}$ has infinite flow property if only if its infinite flow graph G^∞ is connected.*

Proof Note that an undirected graph $([m], \mathcal{E})$ is connected if and only if it does not have an empty cut $[S, \bar{S}]$ for a non-trivial $S \subset [m]$. Also, note that $A_{ij}(k) + A_{ji}(k) \leq A_S(k)$ for any $i \in S$ and $j \in \bar{S}$. Thus, $\{i, j\} \in [S, \bar{S}]$ in the infinite flow graph G^∞ if $\sum_{k=0}^{\infty} A_S(k) = \infty$.

For the reverse implication suppose that $\sum_{k=0}^{\infty} A_S(k) = \infty$, for some non-trivial $S \subset [m]$. Observe that

$$A_S(k) = \sum_{i \in S, j \in \bar{S}} \left(A_{ij}(k) + A_{ji}(k) \right),$$

and the sets S and $\bar{S}$ have finite cardinality. Therefore, we have

$$\sum_{k=0}^{\infty} \left(A_{ij}(k) + A_{ji}(k) \right) = \infty,$$

for some $i \in S$ and $j \in \bar{S}$. But this happens for any non-trivial subset $S \subseteq [m]$, which implies that G^{∞} is a connected graph.   **Q.E.D**

Lemma 3.3 provides an alternative for Theorem 3.1 as follows.

Corollary 3.1 *A deterministic chain $\{A(k)\}$ is ergodic only if its infinite flow graph is connected.*

Thus, if the infinite flow graph is not connected, we cannot have an ergodic chain. Nevertheless, we may have other plausible limiting behavior such as existence of $\lim_{k\to\infty} x(k)$ for any dynamics $\{x(k)\}$ driven by such a chain (that does not have infinite flow property). The next step in our analysis is to characterize properties of the limit points of such dynamics given that such a limit exists.

To continue our discussion, let us consider the following definition.

Definition 3.4 Let $\{A(k)\}$ be a stochastic chain. Then, we say that i,j are mutually ergodic indices for $\{A(k)\}$ and we denote it by $i \leftrightarrow_A j$ if $\lim_{k\to\infty} \left(x_i(k) - x_j(k) \right) = 0$ for any dynamics $\{x(k)\}$ driven by $\{A(k)\}$ started with an arbitrary initial condition $(t_0, x(t_0)) \in \mathbb{Z}^+ \times \mathbb{R}^m$.

We say that $i \in [m]$ is an ergodic index for $\{A(k)\}$ if $\lim_{k\to\infty} x_i(k)$ exists for any initial condition $(t_0, x(t_0)) \in \mathbb{Z}^+ \times \mathbb{R}^m$.

As an example consider an ergodic chain $\{A(k)\}$. Then, $i \leftrightarrow_A j$ for any $i, j \in [m]$. Moreover, any $i \in [m]$ is an ergodic index for such a chain. Using the argument as in the proof of Theorem 2.1, one can show that $i \leftrightarrow_A j$ if and only if the difference between the ith and jth column of $A(k : t_0)$ goes to zero as k approached infinity, for any starting time t_0. Similarly, it can be shown that $i \in [m]$ is an ergodic index if $\lim_{k\to\infty} A_i(k : t_0)$ exists for any starting time $t_0 \geq 0$. So, we have the following result.

Lemma 3.4 *For a chain $\{A(k)\}$, we have $i \leftrightarrow_A j$ if and only if*

$$\lim_{k\to\infty} \left(A_{i\ell}(k : t_0) - A_{j\ell}(k : t_0) \right) = 0,$$

for any $\ell \in [m]$ and any $t_0 \geq 0$. Also, $i \in [m]$ is an ergodic index if and only if $\lim_{k\to\infty} A_i(k : t_0)$ exists for any $t_0 \geq 0$.

In the case of ergodicity, from the mutual ergodicity of all pairs of indices (i.e. weak ergodicity), we can also conclude that every index is an ergodic index. However, mutual ergodicity of two indices, on its own, does not imply that either of the indices is ergodic, as shown in the following example.

Example 3.1 Consider the 4×4 stochastic chain $\{A(k)\}$ defined by:

$$A(2k) = \begin{bmatrix} 1\ 0\ 0\ 0 \\ 1\ 0\ 0\ 0 \\ 1\ 0\ 0\ 0 \\ 0\ 0\ 0\ 1 \end{bmatrix} \text{ and } A(2k+1) = \begin{bmatrix} 1\ 0\ 0\ 0 \\ 0\ 0\ 0\ 1 \\ 0\ 0\ 0\ 1 \\ 0\ 0\ 0\ 1 \end{bmatrix},$$

for any $k \geq 0$. It can be verified that for any starting time $t_0 \geq 0$ and any $k \geq t_0$, we have $A(k : t_0) = A(k)$. This implies that $2 \leftrightarrow_A 3$ while $\lim_{k \to \infty} A_2(k : t_0)$ and $\lim_{k \to \infty} A_3(k : t_0)$ do not exist.

Our goal is to show that a *small* perturbation of any stochastic chain preserves mutual ergodicity and the set of ergodic indices. Let us define precisely what a *small* perturbation is.

Definition 3.5 (ℓ_1-approximation) We say that a chain $\{B(k)\}$ is an ℓ_1-approximation of a chain $\{A(k)\}$ if

$$\sum_{k=0}^{\infty} |A_{ij}(k) - B_{ij}(k)| < \infty \qquad \text{for all } i, j \in [m].$$

Since the set of all absolutely summable sequences in $\mathbb{R}$ is a vector space over $\mathbb{R}$, it follows that ℓ_1-approximation is an equivalence relation for deterministic chains. Second, we note that there are alternative formulations of ℓ_1-approximation. Since the matrices have a finite dimension, we have $\sum_{k=0}^{\infty} |A_{ij}(k) - B_{ij}(k)| < \infty$ for all $i, j \in [m]$ if and only if $\sum_{k=0}^{\infty} \|A(k) - B(k)\|_p < \infty$ for any $p \geq 1$. Thus, an equivalent definition of ℓ_1-approximation is obtained by requiring that $\sum_{k=0}^{\infty} \|A(k) - B(k)\|_p < \infty$ for some $p \geq 1$.

As an example of ℓ_1-approximation, let $\{A(k)\}$ be an arbitrary stochastic chain and let $\{B(k)\}$ be a chain such that $A(k) \neq B(k)$ for finitely many indices $k \geq 0$. Then, $\{A(k)\}$ is an ℓ_1-approximation of $\{B(k)\}$. Note that such an approximation of $\{A(k)\}$ has ergodic properties similar to the ergodic properties of $\{B(k)\}$, i.e. $i \leftrightarrow_A j$ if and only if $i \leftrightarrow_B j$ and, also, $i \in [m]$ is an ergodic index for $\{A(k)\}$ if and only if $i \in [m]$ is an ergodic index for $\{B(k)\}$. In fact, this property holds for any ℓ_1-approximation of any stochastic chain.

Lemma 3.5 (Approximation lemma) *Let a deterministic chain $\{B(k)\}$ be an ℓ_1-approximation of a deterministic chain $\{A(k)\}$. Then, $i \leftrightarrow_A j$ if and only if $i \leftrightarrow_B j$ and $i \in [m]$ is an ergodic index for $\{A(k)\}$ if and only if it is an ergodic index for $\{B(k)\}$.*

Proof Suppose that $i \leftrightarrow_B j$. Let $t_0 = 0$ and let $x(0) \in [0, 1]^m$. Also, let $\{x(k)\}$ be the dynamics driven by $\{A(k)\}$. For any $k \geq 0$, we have

$$x(k+1) = A(k)x(k) = (A(k) - B(k))x(k) + B(k)x(k).$$

Since $|x_i(k)| \leq 1$ for any $k \geq 0$ and any $i \in [m]$, it follows that for all $k \geq 0$,

$$\|x(k+1) - B(k)x(k)\|_\infty \leq \|A(k) - B(k)\|_\infty. \qquad (3.2)$$

We want to show that $i \leftrightarrow_A j$, or equivalently that $\lim_{k\to\infty}(x_i(k) - x_j(k)) = 0$. To do so, we let $\epsilon > 0$ be arbitrary but fixed. Since $\{B(k)\}$ is an ℓ_1-approximation of $\{A(k)\}$, there exists time $N_\epsilon \geq 0$ such that $\sum_{k=N_\epsilon}^{\infty} \|A(k) - B(k)\|_\infty \leq \epsilon$. Let $\{z(k)\}_{k \geq N_\epsilon}$ be the dynamics driven by $\{B(k)\}$ and started at time N_ϵ with the initial point $z(N_\epsilon) = x(N_\epsilon)$. We next show that

$$\|x(k+1) - z(k+1)\|_\infty \leq \sum_{t=N_\epsilon}^{k} \|A(t) - B(t)\|_\infty \qquad \text{for all } k \geq N_\epsilon. \qquad (3.3)$$

We use the induction on k, so we consider $k = N_\epsilon$. Then, by Eq.(3.2), we have $\|x(N_\epsilon + 1) - B(N_\epsilon)x(N_\epsilon)\|_\infty \leq \|A(N_\epsilon) - B(N_\epsilon)\|_\infty$. Since $z(N_\epsilon) = x(N_\epsilon)$, it follows that $\|x(N_\epsilon + 1) - z(N_\epsilon + 1)\|_\infty \leq \|A(N_\epsilon) - B(N_\epsilon)\|_\infty$. We now assume that $\|x(k) - z(k)\|_\infty \leq \sum_{t=N_\epsilon}^{k-1} \|A(t) - B(t)\|_\infty$ for some $k > N_\epsilon$. Using Eq.(3.2) and the triangle inequality, we have

$$\begin{aligned}
\|x(k+1) - z(k+1)\|_\infty &= \|A(k)x(k) - B(k)z(k)\|_\infty \\
&= \|(A(k) - B(k))x(k) + B(k)(x(k) - z(k))\|_\infty \\
&\leq \|(A(k) - B(k))\|_\infty \|x(k)\|_\infty \\
&\quad + \|B(k)\|_\infty \|(x(k) - z(k))\|_\infty.
\end{aligned}$$

By the induction hypothesis and relation $\|B(k)\|_\infty = 1$, which holds since $B(k)$ is a stochastic matrix, it follows that $\|x(k+1) - z(k+1)\|_\infty \leq \sum_{t=N_\epsilon}^{k} \|A(t) - B(t)\|_\infty$, thus showing relation (3.3).

Recalling that the time $N_\epsilon \geq 0$ is such that $\sum_{k=N_\epsilon}^{\infty} \|A(k) - B(k)\|_\infty \leq \epsilon$ and using relation (3.3), we obtain for all $k \geq N_\epsilon$,

$$\|x(k+1) - z(k+1)\|_\infty \leq \sum_{t=N_\epsilon}^{k} \|A(t) - B(t)\|_\infty \leq \sum_{t=N_\epsilon}^{\infty} \|A(t) - B(t)\|_\infty \leq \epsilon. \qquad (3.4)$$

Therefore, $|x_i(k) - z_i(k)| \leq \epsilon$ and $|z_j(k) - x_j(k)| \leq \epsilon$ for any $k \geq N_\epsilon$, and by the triangle inequality we have $|(x_i(k) - x_j(k)) + (z_i(k) - z_j(k))| \leq 2\epsilon$ for any $k \geq N_\epsilon$. Since $i \leftrightarrow_B j$, it follows that $\lim_{k\to\infty}(z_i(k) - z_j(k)) = 0$ and $\limsup_{k\to\infty} |x_i(k) - x_j(k)| \leq 2\epsilon$. The preceding relation holds for any $\epsilon > 0$, implying that $\lim_{k\to\infty}(x_i(k) - x_j(k)) = 0$. Furthermore, the same analysis would go through when t_0 is arbitrary and the initial point $x(0) \in \mathbb{R}^m$ is arbitrary with $\|x(0)\|_\infty \neq 1$. Thus, we have $i \leftrightarrow_A j$.

Using the same argument and inequality (3.4), one can deduce that if i is an ergodic index for $\{B(k)\}$, then it is also an ergodic index for $\{A(k)\}$. Since ℓ_1-approximation is symmetric with respect to the chains, the result follows. **Q.E.D**

The ℓ_1-approximation lemma has many interesting implications. The first implication is that ℓ_1-approximation preserves ergodicity.

Corollary 3.2 *Let $\{B(k)\}$ be an ℓ_1-approximation of a chain $\{A(k)\}$. Then, $\{B(k)\}$ is ergodic if and only if $\{A(k)\}$ is ergodic.*

Proof Note that a chain $\{A(k)\}$ is ergodic if and only if $i \leftrightarrow_A j$ for any $i, j \in [m]$. So if $\{B(k)\}$ is an ℓ_1-approximation of $\{A(k)\}$, then Lemma 3.5 implies that $\{A(k)\}$ is ergodic if and only if $\{B(k)\}$ is ergodic.   **Q.E.D**

Another implication of ℓ_1-approximation lemma is a generalization of Theorem 3.1. Recall that by Theorem 3.1, the ergodicity of a chain $\{A(k)\}$ implies the connectivity of the infinite flow graph of $\{A(k)\}$.

Lemma 3.6 *Let $\{A(k)\}$ be a deterministic chain and let G^∞ be its infinite flow graph. Then, $i \leftrightarrow_A j$ implies that i and j belong to the same connected component of G^∞.*

Proof To arrive at a contradiction, suppose that i and j belong to two different connected components $S, T \subset [m]$ of G^∞. Therefore, $T \subset \bar{S}$ implying that $\bar{S}$ is not empty. Also, since S is a connected component of G^∞, it follows that $\sum_{k=0}^\infty A_S(k) < \infty$. Without loss of generality, we assume that $S = \{1, \ldots, i^*\}$ for some $i^* < m$, and consider the chain $\{B(k)\}$ defined by

$$B_{ij}(k) = \begin{cases} A_{ij}(k) & \text{if } i \neq j \text{ and } i, j \in S \text{ or } i, j \in \bar{S}, \\ 0 & \text{if } i \neq j \text{ and } i \in S, j \in \bar{S} \text{ or } i \in \bar{S}, j \in S, \\ A_{ii}(k) + \sum_{\ell \in \bar{S}} A_{i\ell}(k) & \text{if } i = j \in S, \\ A_{ii}(k) + \sum_{\ell \in S} A_{i\ell}(k) & \text{if } i = j \in \bar{S}. \end{cases} \tag{3.5}$$

The above approximation simply sets the cross terms between S and $\bar{S}$ to zero, and adds the deleted values to the corresponding diagonal entries to maintain the stochasticity of the matrix $B(k)$. Therefore, for the stochastic chain $\{B(k)\}$ we have

$$B(k) = \begin{bmatrix} B_1(k) & 0 \\ 0 & B_2(k) \end{bmatrix},$$

where $B_1(k)$ and $B_2(k)$ are respectively $i^* \times i^*$ and $(m - i^*) \times (m - i^*)$ matrices for all $k \geq 0$. By the assumption $\sum_{k=0}^\infty A_S(k) < \infty$, the chain $\{B(k)\}$ is an ℓ_1-approximation of $\{A(k)\}$. Now, let u_{i^*} be the vector which has the first i^* coordinates equal to one and the rest equal to zero, i.e., $u_{i^*} = \sum_{\ell=1}^{i^*} e_\ell$. Then, $B(k)u_{i^*} = u_{i^*}$ for any $k \geq 0$ implying that $i \nleftrightarrow_B j$. By approximation lemma (Lemma 3.5) it follows $i \nleftrightarrow_A j$, which is a contradiction.   **Q.E.D**

Lemma 3.6 shows that $i \leftrightarrow_A j$ is possible only for indices that fall in the same connected component of the infinite flow graph of $\{A(k)\}$. However, the reverse implication is not true as illustrated by the following example.

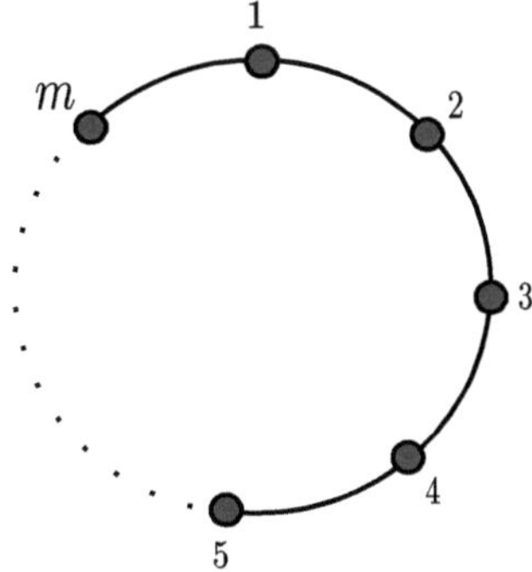

Fig. 3.2 The infinite flow graph of the chain discussed in Example 3.2

Example 3.2 Let $\{A(k)\}$ be the static chain defined by

$$A(k) = A(0) = \begin{bmatrix} 0 & 1 & 0 & \cdots & 0 \\ 0 & 0 & 1 & \cdots & 0 \\ \vdots & \vdots & \vdots & \ddots & \vdots \\ 0 & 0 & 0 & \cdots & 1 \\ 1 & 0 & 0 & \cdots & 0 \end{bmatrix} \qquad \text{for any } k \geq 0.$$

Then, the infinite flow graph of $\{A(k)\}$ is a cycle as shown in Fig. 3.2 and, hence, it is connected. Nevertheless, $A(0)$ is just a permutation matrix and, hence, $i \not\leftrightarrow_A j$ for any $i, j \in [m]$.

Another implication of the ℓ_1-approximation is an extension to Theorem 2.1 where it is shown that weak ergodicity implies strong ergodicity. Note that weak ergodicity of a chain $\{A(k)\}$ is equivalent to $i \leftrightarrow_A j$ for any $i, j \in [m]$. On the other hand, any weak ergodic chain has infinite flow property and hence, any two indices $i, j \in [m]$ belong to a same connected component of the infinite flow graph of $\{A(k)\}$. Based on this observation, the following result extends Theorem 2.1 to non-ergodic chains.

Lemma 3.7 *Suppose that $i \leftrightarrow_A j$ for any $i, j \in S$ where S is a connected component of G^∞. Then, any $i \in S$ is an ergodic index.*

Proof. Without loss of generality let us assume that $S = \{1, \ldots, i^*\}$ for some $i^* \in [m]$. For the given chain $\{A(k)\}$ and the connected component S, let $\{B(k)\}$ be the approximation introduced in Eq.(3.5). Then, we have

$$B(k) = \begin{bmatrix} B_1(k) & 0 \\ 0 & B_2(k) \end{bmatrix},$$

for any $k \geq 0$. As it is shown in the proof of Lemma 3.6, it follows that $\{B(k)\}$ is an ℓ_1-approximation of $\{A(k)\}$. But since $i \leftrightarrow_A j$ for any $i, j \in S$, by Lemma 3.5 it follows that $i \leftrightarrow_B j$ for any $i, j \in S$. But by the block diagonal form of $\{B(k)\}$, this implies that $i \leftrightarrow_{B_1} j$ for any $i, j \in S$. By Theorem 2.1 it follows that any $i \in S$ is an ergodic index for $\{B_1(k)\}$ which implies that any $i \in S$ is an ergodic index for

$\{B(k)\}$. From Lemma 3.5, it follows that any $i \in S$ is an ergodic index for $\{A(k)\}$.
Q.E.D

We can extend the notions of ergodic index and mutual ergodicity to a random chain. As for ergodicity and consensus, in this case, we naturally have *subsets* of the sample space over which those properties hold.

Definition 3.6 For a random chain $\{W(k)\}$, we let $i \leftrightarrow_W j \subseteq \Omega$ be the set over which i and j are mutually ergodic indices. We also let $\mathscr{E}_i \subseteq \Omega$ be the subset over which $i \in [m]$ is an ergodic index for $\{W(k)\}$.

Note that by Definition 3.6 and the definition of ergodicity for random chains we have $\mathscr{E} = \bigcap_{i,j \in [m]} i \leftrightarrow_W j$. Moreover, we have $\mathscr{E} \subseteq \bigcap_{i \in [m]} \mathscr{E}_i$.

Similarly, we can define the infinite flow graph for a random chain $\{W(k)\}$. In this case, instead of a deterministic graph, we would have a random graph associated with the given random chain. For this, recall that $\mathcal{G}([m])$ is the set of all simple graphs on vertex set $[m]$.

Definition 3.7 For an adapted chain $\{W(k)\}$, we define infinite flow graph G^∞ : $\Omega \to \mathcal{G}([m])$ by $G^\infty(\omega) = ([m], \varepsilon^\infty(\omega))$ as the graph with the vertex set $[m]$ and the edge set

$$\varepsilon^\infty(\omega) = \{\{i, j\} \mid \sum_{k=0}^{\infty} \left(W_{ij}(k, \omega) + W_{ji}(k, \omega)\right) = \infty, i \neq j \in [m]\}.$$

Using the same lines of argument as in the proofs of Lemma 3.1 and Lemma 3.2, one can show the following result.

Lemma 3.8 *Let G^∞ be the infinite flow random graph associated with a random chain $\{W(k)\}$ in a probability space $(\Omega, \mathcal{F})$. Then, the set $G^{\infty-1}(G) = \{\omega \mid G^\infty(\omega) = G\} \in \mathcal{F}$ for any $G \in \mathcal{G}([m])$.*

Note that the set of events $\{G^{\infty-1}(G) \mid G \in \mathcal{G}([m])\}$ is a partition of the probability space Ω.

Lemma 3.6 naturally holds for random chains as provided below.

Lemma 3.9 *Let $\{W(k)\}$ be a random chain. Then, $i \leftrightarrow_W j \subseteq \Theta_{ij}$ where Θ_{ij} is the event that i and j belong to the same connected component of the infinite flow graph of $\{W(k)\}$.*

Chapter 4
Infinite Flow Stability

In Chap. 3, we showed that regardless of the structure and any assumption on a random chain we have $i \leftrightarrow_W j \subseteq \Theta_{ij}$ for any $i,\, j \in [m]$ where Θ_{ij} is the event that i, j belong to the same connected component of the infinite flow graph of $\{W(k)\}$. In this chapter, we characterize a class of random chains $\{W(k)\}$ for which $i \leftrightarrow_W j = \Theta_{ij}$ almost surely, i.e. we can characterize the limiting behavior of the random dynamics (3.1) (or the product of random stochastic matrices) by inspecting the infinite flow graph of the given random chain.

The structure of this chapter is as follows: in Sect. 4.1 we introduce the concept of the infinite flow stability which is the central notion of this chapter. In Sect. 4.2, we introduce one of the main properties that is needed to ensure the infinite flow stability, i.e. several notions of feedback properties, and we study the relations of the different types of feedback properties with each other. In Sect. 4.3, we introduce an infinite family of comparison functions for the averaging dynamics including a quadratic one. We also prove a fundamental relation that provides a bound for the decrease of the quadratic comparison function. In Sect. 4.4, we characterize a class of random chains, the class $\mathcal{P}^*$, and we prove one of the central results of this thesis which states that any chain in class $\mathcal{P}^*$ with a proper feedback property is infinite flow stable. Finally, in Sect. 4.5, we introduce a class of random chains that are more of practical interest and belong to the class $\mathcal{P}^*$. We show that many of the known ergodic (and hence, infinite flow stable) chains are members of this class.

4.1 Infinite Flow Stability

As shown in Lemma 3.6, for a stochastic chain $\{A(k)\}$, two indices can be mutually ergodic only if they belong to the same connected component of the infinite flow graph of $\{A(k)\}$. This result provides the minimum requirement for mutual ergodicity and ergodicity. However, as discussed in Example 3.1, this condition is not necessary, i.e. in general we cannot conclude mutual ergodicity by inspecting the infinite flow

B. Touri, *Product of Random Stochastic Matrices and Distributed Averaging*,
Springer Theses, DOI: 10.1007/978-3-642-28003-0_4,
© Springer-Verlag Berlin Heidelberg 2012

graph of a chain. We are interested in characterization of chains for which the reverse implication holds which are termed as the *infinite flow stable* chains.

Definition 4.1 (*Infinite Flow Stability*) We say that a chain $\{A(k)\}$ is infinite flow stable if the dynamics $\{x(k)\}$ converges for any initial condition $(t_0, x(t_0)) \in \mathbb{Z}^+ \times \mathbb{R}^m$ and $\lim_{k\to\infty} (x_i(k) - x_j(k)) = 0$ for all $\{i, j\} \in \mathcal{E}^\infty$, where $\mathcal{E}^\infty$ is the edge set of the infinite flow graph of $\{A(k)\}$.

As in the case of the ergodicity and consensus (Theorem 2.2 and 2.3), the infinite flow stability of a chain can be equivalently characterized by the product of stochastic matrices in the given chain.

Lemma 4.1 *A stochastic chain $\{A(k)\}$ is infinite flow stable if and only if $\lim_{k\to\infty} A(k : t_0)$ exists for all $t_0 \geq 0$ and also $\lim_{k\to\infty} \|A_i(k : t_0) - A_j(k : t_0)\| = 0$ for any i, j belonging to the same connected component of the infinite flow graph G^∞ of $\{A(k)\}$.*

Proof Note that by Lemma 3.7, the infinite flow stability implies that any dynamics $\{x(k)\}$ driven by $\{A(k)\}$ is convergent. Thus, if we let $x_i(t_0) = e_\ell$, it follows that $\lim_{k\to\infty} A^\ell(k : t_0)$ exists, and since this holds for any $x(t_0) = e_\ell$ it follows that $\lim_{k\to\infty} A(k : t_0)$ exists for any $t_0 \geq 0$. Moreover, since for any i, j in the same connected component of G^∞ we have $\lim_{k\to\infty} (x_i(k) - x_j(k)) = 0$, it follows that $\lim_{k\to\infty} (A_{i\ell}(k : t_0) - A_{j\ell}(k : t_0)) = 0$ for any $\ell \in [m]$. Thus, $\lim_{k\to\infty} \|A_i(k : t_0) - A_j(k : t_0)\| = 0$ for any $t_0 \geq 0$ and any i, j in the same connected component of G^∞.

For the converse, note that for any $(t_0, x(t_0)) \in \mathbb{Z}^+ \times \mathbb{R}^m$ and any $k \geq t_0$, we have

$$|x_i(t_0) - x_j(t_0)| = |(A_i(k : t_0) - A_j(k : t_0)) x(t_0)| \leq \|A_i(k : t_0) - A_j(k : t_0)\| \|x(t_0)\|,$$

which follows by Cauchy-Schwartz inequality. Thus, if $\lim_{k\to\infty} \|A_i(k : t_0) - A_j(k : t_0)\| = 0$, for any $x(t_0) \in \mathbb{R}^m$, we have $\lim_{k\to\infty} (x_i(t_0) - x_j(t_0)) = 0$.   **Q.E.D.**

The immediate question is that if for some chain $\{A(k)\}$ the product $A(k : t_0)$ is convergent for any $t_0 \geq 0$, can we conclude the infinite flow stability of $\{A(k)\}$? In other words, is there any chain $\{A(k)\}$ such that any dynamics driven by $\{A(k)\}$ is convergent for any initial condition $(t_0, x(t_0)) \in \mathbb{Z}^+ \times \mathbb{R}^m$ but yet, $\{A(k)\}$ is not infinite flow stable? The following example shows that infinite flow stability is in fact stronger than asymptotic stability.

Example 4.1 Consider the stochastic chain $\{A(k)\}$ in $\mathbb{R}^{3\times3}$ defined by $A(0) = I$ and

$$A(k) = \begin{bmatrix} 1 & 0 & 0 \\ 1 - \frac{1}{k} & 0 & \frac{1}{k} \\ 0 & 0 & 1 \end{bmatrix} \quad \text{for} \quad k \geq 1.$$

In this case, it can be verified that $A(k : t_0) = A(k)$ for any $t_0 \geq 0$ and $k > t_0$. Therefore, we have

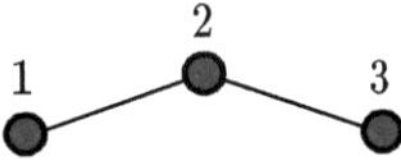

Fig. 4.1 The infinite flow graph of the chain discussed in Example 4.1

$$\lim_{k \to \infty} A(k : t_0) = \lim_{k \to \infty} A(k) = \begin{bmatrix} 1 & 0 & 0 \\ 1 & 0 & 0 \\ 0 & 0 & 1 \end{bmatrix}.$$

Note that the infinite flow graph of the chain $\{A(k)\}$ is connected (see Fig. 4.1). However, the rows of the limiting matrix are not the same (i.e. $\{A(k)\}$ is not ergodic). Therefore, in spite of convergence of $A(k : t_0)$ for any $t_0 \geq 0$, $\{A(k)\}$ is not infinite flow stable.

We can extend the notion of infinite flow stability to a random chain $\{W(k)\}$. For a random chain $\{W(k)\}$, the infinite flow stability happens on a measurable subset of Ω. We say that a chain $\{W(k)\}$ is infinite flow stable if it is infinite flow stable almost surely.

4.2 Feedback Properties

As in the definition of B-connected chains (Definition 2.4 (b)), in order to ensure the convergence of the dynamics (2.2), it is often assumed that diagonal entries of the matrices in a stochastic chain $\{A(k)\}$ are uniformly bounded from below by a constant scalar. From the opinion dynamics viewpoint, such an assumption can be considered as a form of self confidence of each agents. It also ensures that effect of the other agents' opinion does not vanish in finite time. Here, we define several notions of feedback property and we discuss some general results for them which will be useful in our further development.

We start our discussion on feedback properties by introducing different types of feedback property which will be considered in this work. We discuss the feedback property in the context of an adapted random chain.

Definition 4.2 (*Feedback Properties*) We say that an adapted random chain $\{W(k)\}$ has:
(a) strong feedback property: if

$$W_{ii}(k) \geq \gamma \qquad \text{almost surely,}$$

for some $\gamma > 0$, any $k \geq 0$, and all $i \in [m]$.
(b) feedback property: if

$$\mathsf{E}\left[W_{ii}(k) W_{ij}(k) \mid \mathcal{F}_k \right] \geq \gamma \mathsf{E}\left[W_{ij}(k) \mid \mathcal{F}_k \right] \qquad \text{almost surely,}$$

for some $\gamma > 0$, and any $k \geq 0$ and all distinct $i, j \in [m]$.
(c) weak feedback property: if

$$\mathsf{E}\left[W^{iT}(k)W^j(k) \mid F_k\right] \geq \gamma \mathsf{E}\left[W_{ij}(k) + W_{ji}(k) \mid \mathcal{F}_k\right] \quad \text{almost surely,}$$

for some $\gamma > 0$, and any $k \geq 0$ and all distinct $i, j \in [m]$.

We refer to γ as the feedback coefficient and without loss of generality we may assume that $\gamma \leq 1$.

In words, weak feedback property requires that the correlation of the ith and jth columns of $W(k)$ is bounded below by a constant factor of $\mathsf{E}\left[W_{ij}(k) + W_{ji}(k) \mid \mathcal{F}_k\right]$ for any $k \geq 0$ and any $i, j \in [m]$. Similarly, the feedback property requires that the correlation of the $W_{ii}(k)W_{ij}(k)$ is bounded below by a constant factor of $\bar{W}_{ij}(k)$.

Note that if $W_{ii}(k) \geq \gamma$ almost surely, then $\mathsf{E}\left[W_{ii}(k)W_{ij}(k) \mid \mathcal{F}_k\right] \geq \gamma \mathsf{E}\left[W_{ij}(k) \mid \mathcal{F}_k\right]$ almost surely and hence, strong feedback property implies feedback property. Also, note that since $W(k)$ s are non-negative almost surely, we have $W^{iT}(k)W^j(k) \geq W_{ii}(k)W_{ij}(k) + W_{jj}(k)W_{ji}(k)$ which shows that the feedback property implies weak feedback property. In this work, most of our main results are based on weak feedback property and feedback property.

We first show that the feedback property implies strong feedback property for the expected chain.

Lemma 4.2 *Let a random chain $\{W(k)\}$ have feedback property with feedback coefficient $\gamma \leq 1$. Then, the expected chain $\{\bar{W}(k)\}$ has strong feedback property with $\frac{\gamma}{1-\gamma}$.*

Proof Let the random chain have feedback property with constant $\gamma > 0$. Then, by the definition of the feedback property, for any time $k \geq 0$ and any $i, j \in [m]$ with $i \neq j$, we have

$$\mathsf{E}\left[W_{ii}(k)W_{ij}(k) \mid \mathcal{F}_k\right] \geq \gamma \, \mathsf{E}\left[W_{ij}(k) \mid \mathcal{F}_k\right] = \gamma \, \bar{W}_{ij}(k).$$

For a fixed $i \in [m]$, by adding up both sides of the above relation over all $j \neq i$, and using the fact that $W(k)$ is stochastic almost surely, we have

$$\mathsf{E}\left[W_{ii}(k)(1 - W_{ii}(k)) \mid \mathcal{F}_k\right] \geq \gamma(1 - \bar{W}_{ii}(k)).$$

But $W_{ii}(k) \leq 1$ almost surely implying $\mathsf{E}\left[W_{ii}(k)(1 - W_{ii}(k)) \mid \mathcal{F}_k\right] \leq \bar{W}_{ii}(k)$. Therefore, we have $\bar{W}_{ii}(k) \geq \gamma(1 - \bar{W}_{ii}(k))$, implying $\bar{W}_{ii}(k) \geq \frac{\gamma}{1-\gamma}$.   **Q.E.D.**

Since for deterministic chains, the expected chain and the original chain are the same, Lemma (4.2) implies that the feedback property and the strong feedback property are equivalent for deterministic chains. However, the following result shows that even for deterministic chains, the feedback property cannot be implied by weak feedback property.

Example 4.2 Consider the static deterministic chain $\{A(k)\}$ given by:

$$A(k) = A = \begin{bmatrix} 0 & \frac{1}{2} & \frac{1}{2} \\ \frac{1}{2} & 0 & \frac{1}{2} \\ \frac{1}{2} & \frac{1}{2} & 0 \end{bmatrix} \qquad \text{for } k \geq 0.$$

Since $A_{ii} A_{ij} = 0$ and $A_{ij} \neq 0$ for all $i \neq j$, the chain does not have feedback property. At the same time, since $A_{ij} + A_{ji} = 1$ for $i \neq j$, it follows $(A^i)^T A^j = \frac{1}{4} = \frac{1}{4}(A_{ij} + A_{ji})$. Thus, $\{A(k)\}$ has weak feedback property with $\gamma = \frac{1}{4}$. $\qquad\square$

Our next goal is to show that i.i.d. chains with almost sure positive diagonal entries have feedback property. To do this, let us first prove the following intermediate result.

Lemma 4.3 *Consider an adapted random chain $\{W(k)\}$. Suppose that the random chain is such that there is an $\eta > 0$ with the following property:*
for all $k \geq 0$ and $i, j \in [m]$ with $i \neq j$,

$$\mathsf{E}\big[W_{ii}(k) W_{ij}(k) \mid \mathcal{F}_k \big] \geq \eta \mathbf{1}_{\{\bar{W}_{ij}(k)>0\}},$$

or the following property:
for all $k \geq 0$ and $i, j \in [m]$ with $i \neq j$,

$$\mathsf{E}\big[W^{iT}(k) W^j(k) \mid \mathcal{F}_k \big] \geq \eta \mathbf{1}_{\{\bar{W}_{ij}(k)>0\}}.$$

Then, respectively, the chain has feedback property with constant η or weak feedback property with constant $\eta/2$.

Proof To prove the case of feedback property, note that $\mathbf{1}_{\bar{W}_{ij}(k)>0} \geq \bar{W}_{ij}(k)$ for any $k \geq 0$ and all distinct $i, j \in [m]$. Thus, $\mathsf{E}\big[W_{ii}(k) W_{ij}(k) \mid \mathcal{F}_k \big] \geq \eta \mathbf{1}_{\bar{W}_{ij}(k)>0} \geq \eta \bar{W}_{ij}(k)$, which implies that $\{W(k)\}$ has feedback property. The case of weak feedback property follows by the same line of argument. **Q.E.D.**

The i.i.d. chains with almost surely positive diagonal entries have been studied in [1–3]. These chains have feedback property, as seen in the following corollary.

Corollary 4.1 *Let $\{W(k)\}$ be an i.i.d. random chain with almost sure positive diagonal entries, i.e. $W_{ii}(k) > 0$ almost surely. Then, $\{W(k)\}$ has feedback property with constant $\gamma = \min_{\{i \neq j \mid \bar{W}_{ij}(k)>0\}} \mathsf{E}\big[W_{ii}(k) W_{ij}(k) \big]$.*

Proof Suppose that $\bar{W}_{ij}(k) > 0$ for some $i, j \in [m]$. Since $W_{ii}(k) > 0$ *a.s.* and $W_{ij}(k) \geq 0$, it follows that $\mathsf{E}\big[W_{ii}(k) W_{ij}(k) > 0 \big]$. Since $\{W(k)\}$ is assumed to be i.i.d., the constant $\eta = \min_{\{i \neq j \mid \bar{W}_{ij}(k)>0\}} \mathsf{E}\big[W_{ii}(k) W_{ij}(k) \big]$ is independent of time. Also, since the index set $i \neq j \in [m]$ is finite, it follows that $\eta > 0$. Hence, by Lemma 4.3 it follows that the chain has feedback property with constant η. **Q.E.D.**

4.3 Comparison Functions for Averaging Dynamics

Here, we introduce an infinite family of comparison functions for random averaging dynamics. To do so, first we will reintroduce the concept of an absolute probability sequence for stochastic chains and then we will introduce an absolute probability process which is the generalization of absolute probability sequences to adapted processes. Using an absolute probability process, and *any* given convex function, we introduce a comparison function for averaging dynamics.

4.3.1 Absolute Probability Process

In [4], Kolmogorov introduced and studied an elegant object, an *absolute probability sequence*, which is a sequence of stochastic vectors associated with a chain of stochastic matrices as defined below.

Definition 4.3 A sequence of stochastic vectors $\{\pi(k)\}$ is said to be an absolute probability sequence for a (deterministic) stochastic chain $\{A(k)\}$, if

$$\pi^T(k+1)A(k) = \pi^T(k) \qquad \text{for all } k \geq 0. \tag{4.1}$$

As an example, let $\{A(k)\}$ be a chain of doubly stochastic matrices. Then, the static sequence $\{\pi(k)\}$ defined by $\pi(k) = \frac{1}{m}e$ for all $k \geq 0$ is an absolute probability sequence for $\{A(k)\}$. As another example, consider the static stochastic chain $\{A(k)\}$ defined by $A(k) = A$ for some stochastic matrix A. Since A is a stochastic matrix, it has a stochastic left-eigenvector π. In this case, the static sequence $\{\pi(k)\}$ defined by $\pi(k) = \pi$ for all $k \geq 0$ is an absolute probability sequence for $\{A(k)\}$. A more non-trivial and interesting example of absolute probability sequences is provided in the following result.

Lemma 4.4 [4] *Let $\{A(k)\}$ be an ergodic chain. Suppose that $\lim_{t \to \infty} A(t:k) = e\pi^T(k)$ for all $k \geq 0$. Then, the sequence $\{\pi(k)\}$ is an absolute probability sequence for $\{A(k)\}$.*

Proof For any $k \geq 0$, we have:

$$e\pi^T(k) = \lim_{t \to \infty} A(t:k) = \lim_{t \to \infty} A(t:k+1)A(k) = e\pi^T(k+1)A(k).$$

Thus multiplying both sides of the above equation by $\frac{1}{m}e^T$, it follows that $\pi^T(k) = \pi^T(k+1)A(k)$ and hence, $\{\pi(k)\}$ is an absolute probability sequence for $\{A(k)\}$. **Q.E.D.**

Note that if we have a vector $\pi(k)$ for some $k > 0$, we can find vectors $\pi(k-1), \ldots, \pi(0)$ that fit Eq. (4.1). However, there is no certificate that we can find vectors $\ldots, \pi(k+2), \pi(k+1)$ that satisfy Eq. (4.1). In general, to ensure the existence of such a sequence, we need to solve an infinite system of linear equations.

The following result from [5] plays an important role in establishing the existence of an absolute probability sequence follows immediately. A proof of this result is presented in [5].

Theorem 4.1 *For any chain of stochastic matrices $\{A(k)\}$, there exists an increasing sequence of integers $\{r_t\}$ such that the limit*

$$\lim_{t \to \infty} A(r_t : k) = Q(k) \tag{4.2}$$

exists for any $k \geq 0$.

By Theorem 4.1, the existence of an absolute probability sequence for any stochastic chain follows immediately.

Theorem 4.2 [5] *Any chain $\{A(k)\}$ of stochastic matrices admits an absolute probability sequence.*

Proof By Theorem 4.1, there exists a subsequence $\{r_t\}$ of non-negative integers such that $Q(k) = \lim_{t \to \infty} A(r_t : k)$ exists. Now, for any $s > k$, we have $Q(s)A(s : k) = Q(k)$. This simply follows from the fact that

$$Q(k) = \lim_{t \to \infty} A(r_t : k) = \left(\lim_{t \to \infty} A(r_t : s) \right) A(s : k) = Q(s)A(s : k).$$

In particular $Q(k+1)A(k+1) = Q(k)$ for any $k \geq 0$. Therefore, for any stochastic $\pi \in \mathbb{R}^m$, the sequence $\{Q^T(k)\pi\}$ is an absolute probability sequence for $\{A(k)\}$. **Q.E.D.**

Theorem 4.2, also can be proven directly using Tychonoff theorem and Tychonoff Fixed Point theorem.

Theorem 4.3 *Any chain $\{A(k)\}$ of stochastic matrices admits an absolute probability sequence.*

Proof Let $\mathbf{V}_m$ be the set of stochastic vectors in $\mathbb{R}^m$ and let $\mathbf{V}_m^{\infty} = \cdots \times \mathbf{V}_m \times \mathbf{V}_m$. Note that $\mathbf{V}_m^{\infty} \subset \cdots \times \mathbb{R}^m \times \mathbb{R}^m = \mathbb{R}^{m \times \infty}$ and $\mathbb{R}^{m \times \infty}$ is a locally convex topological vector space (with respect to the product topology of the standard topology on $\mathbb{R}^m$, i.e., the topology generated by the open balls in $\mathbb{R}^m$). Since $\mathbf{V}_m$ is compact, by Tychonoff's Theorem B.2, it follows that $\mathbf{V}_m^{\infty}$ is also a compact set (with respect to the product topology).

Now, for any stochastic chain $A = \{A(k)\}$, let the transformation $T_A : \mathbf{V}_m^{\infty} \to \mathbf{V}_m^{\infty}$ be defined by

$$T_A(\{\pi(k)\})(t) = \pi^T(t+1)A(t) \qquad \text{for } t \geq 0,$$

where $T_A(\{\pi(k)\})(t)$ is the t'th coordinate of the chain $T_A(\{\pi(k)\})$. Since each coordinate map is continuous, it can be verified that T_A is a continuous mapping from $\mathbf{V}_m^{\infty}$ to $\mathbf{V}_m^{\infty}$. Thus, by Tychonoff's Fixed Point Theorem B.3, T_A admits a fixed point, i.e. there exists $\{\pi(k)\} \in \mathbf{V}_m^{\infty}$ such that $T_A(\{\pi(k)\}) = \{\pi(k)\}$. Thus, the claim follows from the fact that an absolute probability vector for $\{A(k)\}$ is nothing but a fixed point of the transformation T_A. **Q.E.D.**

By Theorem 4.2, for any random chain $\{W(k)\}$ and for any $\omega \in \Omega$, the chain $\{W(k, \omega)\}$ admits an absolute probability sequence $\{\pi(k, \omega)\}$.

We use the following generalization of absolute probability sequences for adapted random chains.

Definition 4.4 A random vector process $\{\pi(k)\}$ is an absolute probability process for $\{W(k)\}$ if we have:

$$\mathsf{E}\left[\pi^T(k+1)W(k) \mid \mathcal{F}_k\right] = \pi^T(k) \qquad \text{for all } k \geq 0,$$

and $\pi(k)$ is a stochastic vector almost surely for any $k \geq 0$.

Note that this definition implies that $\{\pi(k)\}$ is adapted to $\{\mathcal{F}_k\}$, i.e. $\pi(k)$ is measurable with respect to $\mathcal{F}_k$.

Note that if we have an independent chain $\{W(k)\}$, if we let $\{\mathcal{F}_k\}$ to be the natural filtration of $\{W(k)\}$, i.e. $\mathcal{F}_k = \sigma(W(k-1), \ldots, W(0))$ and $\mathcal{F}_0 = \{\Omega, \emptyset\}$, then $\bar{W}(k) = \mathsf{E}\left[W(k) \mid \mathcal{F}_k\right]$ would be a deterministic matrix for any $k \geq 0$. Thus, in this case an absolute probability sequence for $\{\bar{W}(k)\}$ is an absolute probability process for $\{W(k)\}$. Thus, by Theorem 4.2, the existence of an absolute probability process for an independent chain follows.

4.3.2 A Family of Comparison Functions

Existence of a quadratic Lyapunov function for averaging dynamics may lead to fast rate of convergence results as well as better understanding of averaging dynamics. As an example, in [6], using a quadratic Lyapunov function a fast rate of convergence has been shown for averaging dynamics driven by a class of doubly stochastic matrices. The method used in [6] appears to be generalizable to any dynamics admitting a quadratic Lyapunov function. Also, as it is pointed out in [7], the existence of such a Lyapunov function may lead to an alternative stability analysis of the dynamics driven by stochastic matrices. However, in [7], the non-existence of a quadratic Lyapunov function for deterministic averaging dynamics is numerically verified for an ergodic deterministic chain. Later, in [8], it is proven analytically that not all the averaging dynamics admit a quadratic Lyapunov function.

Although averaging dynamics may not admit a quadratic Lyapunov function, in this section, we show that *any* dynamics driven by a stochastic chain admits infinitely many comparison functions among which there exists a quadratic one. We furthermore show that the fundamental relation that was essential to derivation of the fast rate of convergence in [6], also holds for a quadratic comparison function.

For this, let $g : \mathbb{R} \to \mathbb{R}$ be an *arbitrary convex function* and let $\pi = \{\pi(k)\}$ be a sequence of stochastic vectors. Let us define $V_{g,\pi} : \mathbb{R}^m \times \mathbb{Z}^+ \to \mathbb{R}^+$ by

$$V_{g,\pi}(x,k) = \sum_{i=1}^{m} \pi_i(k)g(x_i) - g(\pi^T(k)x). \qquad (4.3)$$

Note that since g is assumed to be a convex function over $\mathbb{R}$, it is continuous and hence, $V_{g,\pi}(x,k)$ is a continuous function of x for any $k \geq 0$. Also, since $\pi(k)$ is a stochastic vector, we have $\sum_{i=1}^{m} \pi_i(k)g(x_i) \geq g(\pi^T(k)x)$ and, hence, $V_{g,\pi}(x,k) \geq 0$ for any $x \in \mathbb{R}^m$ and $k \geq 0$. Moreover, if $\pi(k) > 0$ and g is strictly convex, then $V_{g,\pi}(x,k) = 0$ if and only if $x_i = \pi^T(k)x$ for all $i \in [m]$, which holds if and only if x belongs to the consensus subspace $\mathbf{C}$. Therefore, if g is strictly convex and $\{\pi(k)\} > 0$, i.e. $\pi_i(k) > 0$ for all $k \geq 0$ and $i \in [m]$, then $V_{g,\pi}(x,k) = 0$ for some $k \geq 0$ if and only if x is an equilibrium of the dynamics (2.2).

Also, since g is a convex function defined on $\mathbb{R}$, it has a sub-gradient $\nabla g(x)$ at each point $x \in \mathbb{R}$. Furthermore, we have

$$\sum_{i=1}^{m} \pi_i(k)\nabla g(\pi^T(k)x)(x_i - \pi^T(k)x) = \nabla g(\pi^T(k)x) \sum_{i=1}^{m} \pi_i(k)(x_i - \pi^T(k)x) = 0.$$

Therefore, since $\pi(k)$ is stochastic, we obtain

$$\begin{aligned}
V_{g,\pi}(x,k) &= \sum_{i=1}^{m} \pi_i(k)\left(g(x_i) - g(\pi^T(k)x)\right) \\
&= \sum_{i=1}^{m} \pi_i(k)\left(g(x_i) - g(\pi^T(k)x) - \nabla g(\pi^T(k)x)(x_i - \pi^T(k)x)\right) \\
&= \sum_{i=1}^{m} \pi_i(k)D_g(x_i, \pi^T(k)x), \qquad (4.4)
\end{aligned}$$

where $D_g(\alpha, \beta) = g(\alpha) - g(\beta) - \nabla g(\beta)(\alpha - \beta)$. If g is a strongly convex function, $D_g(\alpha, \beta)$ is the Bregman divergence (distance) of α and β with respect to the convex function $g(\cdot)$ as defined in [9]. Equation (4.4) shows that our comparison function is in fact a weighted average of the Bregman divergence of the points $x_1, \ldots, x_m$ with respect to their time changing weighted center of the mass $\pi^T(k)x$.

We now show that $V_{g,\pi}$ is a comparison function for the random dynamics in (3.1).

Lemma 4.5 *For dynamics $\{x(k)\}$ driven by an adapted random chain $\{W(k)\}$ with an absolute probability process $\{\pi(k)\}$, we have almost surely*

$$\mathsf{E}\left[V_{g,\pi}(x(k+1), k+1) \mid \mathcal{F}_k\right] \leq V_{g,\pi}(x(k), k) \qquad \text{for any } k \geq 0.$$

Proof By the definition of $V_{g,\pi}$ in Eq. (4.3), we have almost surely

$$
\begin{aligned}
V_{g,\pi}(x(k+1), k+1) &= \sum_{i=1}^{m} \pi_i(k+1)g(x_i(k+1)) - g(\pi^T(k+1)x(k+1)) \\
&= \sum_{i=1}^{m} \pi_i(k+1)g([W(k)x(k)]_i) - g(\pi^T(k+1)x(k+1)) \\
&\le \sum_{i=1}^{m} \pi_i(k+1) \sum_{j=1}^{m} W_{ij}(k)g(x_j(k)) - g(\pi^T(k+1)x(k+1)),
\end{aligned}
$$

$$(4.5)$$

where the inequality follows by convexity of $g(\cdot)$ and $W(k)$ being stochastic almost surely. Now, since $\{\pi(k)\}$ is an absolute probability process, we have $\mathsf{E}[\pi^T(k+1)W(k) \mid \mathcal{F}_k] = \pi^T(k)$. Also, $x(k)$ is measurable with respect to $\mathcal{F}_k$ and hence, by taking the conditional expectation with respect to $\mathcal{F}_k$ on both sides of Eq. (4.5), we have almost surely

$$
\begin{aligned}
\mathsf{E}\big[V_{g,\pi}(x(k+1), k+1) \mid \mathcal{F}_k\big] &\le \sum_{j=1}^{m} \pi_j(k)g(x_j(k)) - \mathsf{E}\big[g(\pi^T(k+1)x(k+1)) \mid \mathcal{F}_k\big] \\
&\le \sum_{j=1}^{m} \pi_j(k)g(x_j(k)) - g\left(\mathsf{E}\big[\pi^T(k+1)x(k+1) \mid \mathcal{F}_k\big]\right),
\end{aligned}
$$

where the last inequality follows by convexity of g and Jensen's inequality (Lemma A.2). Using $x(k+1) = W(k)x(k)$ and the definition of absolute probability process (Definition 4.4), we obtain almost surely

$$
\mathsf{E}\big[V_{g,\pi}(x(k+1), k+1) \mid F_k\big] \le \sum_{j=1}^{m} \pi_j(k)g(x_j(k)) - g(\pi^T(k)x(k)) = V_{g,\pi}(x(k), k).
$$

Q.E.D.

Lemma 4.5 proves that any adapted random dynamics with an absolute probability process admits infinitely many comparison functions. In particular, any independent and deterministic averaging dynamics admits infinitely many comparison functions through the use of any absolute probability sequence and any convex function g .

Here, we mention two particular choices of the convex function g for constructing comparison functions that might be of particular interest:

1. *Quadratic function*: Let $g(s) = s^2$ and let us consider the formulation provided in Eq. (4.4). Then, it can be seen that

$$
V_{g,\pi}(x, k) = \sum_{i=1}^{m} \pi_i(k)(x_i - \pi^T(k)x)^2.
\tag{4.6}
$$

Thus for any dynamics $\{x(k)\}$ in (3.1), the sequence $\{V_{g,\pi}(x(k), k)\}$ is a non-negative super-martingale in $\mathbb{R}$ and, hence, by Corollary A.3, the sequence $\{V_{g,\pi}(x(k), k)\}$ is convergent almost surely.

2. *Kullback-Leibler divergence*: Let $x(0) \in [0, 1]^m$. One can view $x(0)$ as a vector of positions of m particles in $[0, 1]$. Intuitively, by the successive weighted averaging of the m particles, the entropy of such system should not increase. Mathematically, this corresponds to the choice of $g(s) = s \ln(s)$ in Eq. (4.3) (with $g(0) = 0$). Then, it can be seen that

$$V_{g,\pi}(x, k) = \sum_{i=1}^{m} \pi_i(k) D_{KL}(x_i, \pi^T(k)x),$$

where $D_{KL}(\alpha, \beta) = \alpha \ln(\frac{\alpha}{\beta})$ is the Kullback–Leibler divergence of α and β.

4.3.3 Quadratic Comparison Functions

For the study of the dynamic system (2.2), the Lyapunov function $d(x) = \max_i x_i - \min_i x_i$ has been often used to prove convergence results and develop rate of convergence result. However, this Lyapunov function often results in a poor rate of convergence for the dynamic (2.2). Using a quadratic Lyapunov function for doubly stochastic chains, a fast rate of convergence has been developed to study the dynamics (2.2). The study is highly dependent on the estimate of decrease of the quadratic Lyapunov function along the trajectories of the dynamics (2.2). In this section, we develop a bound for the decrease rate of the quadratic comparison function along the trajectories of (2.2).

In our further development, we consider a *quadratic comparison function* as in (4.6). To simplify the notation, we omit the subscript g in $V_{g,\pi}$ and define

$$V_\pi(x, k) = \sum_{i=1}^{m} \pi_i(k)(x_i - \pi^T(k)x)^2. \tag{4.7}$$

We refer to $V_\pi(x, k)$ as *the quadratic comparison function associated with* $\{\pi(k)\}$. By Theorem 4.2, the existence of such a comparison function follows immediately.

Corollary 4.2 *Every independent random chain* $\{W(k)\}$ *admits a quadratic comparison function.*

In particular, the result of Corollary 4.2 holds for any deterministic chain $\{A(k)\}$.

Some Geometrical Insights

Let us discuss some geometric aspects of the introduced quadratic comparison function (4.7). For a (deterministic) stochastic vector π, consider the weighted semi-norm[1] of two vectors in $\mathbb{R}^m$ defined by $\|x - y\|_\pi^2 = \sum_{i=1}^{m} \pi_i(x_i - y_i)^2$. For any

---

[1] We refer to $\| \cdot \|_\pi$ as a semi-norm because in general $\|x\|_\pi = 0$ does not imply $x = 0$, unless $\pi > 0$ in which case $\| \cdot \|_\pi$ is a norm.

Fig. 4.2 The vector $(\pi^T x)e$ is the projection of a point $x \in \mathbb{R}^m$ on the consensus line **C** in $\|\cdot\|_\pi$ semi-distance

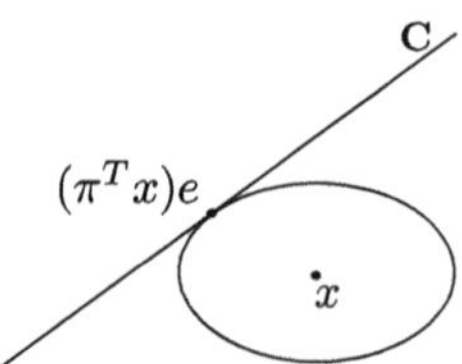

arbitrary point $x \in \mathbb{R}^m$ and an arbitrary point λe in the consensus subspace **C**, we have $\|x - \lambda e\|_\pi^2 = \sum_{i=1}^m \pi_i (x_i - \lambda)^2$. Note that $\lambda^* = \sum_{i=1}^m \pi_i x_i = \pi^T x$ minimizes $\|x - \lambda e\|_\pi^2$ as a function of $\lambda \in \mathbb{R}$. Therefore, the following result holds.

Lemma 4.6 *For an arbitrary point $x \in \mathbb{R}^m$ and a stochastic vector π, the scalar $\pi^T x$ is the projection of the point x on the consensus subspace **C**. In other words,*

$$\|x - (\pi^T x)e\|_\pi^2 = \sum_{i=1}^m \pi_i (x_i - \pi^T x)^2 \le \sum_{i=1}^m \pi_i (x_i - c)^2 = \|x - ce\|_\pi^2,$$

for any $c \in \mathbb{R}$.

This fact is illustrated in Fig. 4.2.

Lemma 4.5 shows that for a dynamics $\{x(k)\}$ driven by a deterministic chain $\{A(k)\}$ at each time instance $k \ge 0$, the distance of the point $x(k)$ with respect to $\|\cdot\|_{\pi(k)}$ semi-norm is non-increasing function of k. Note that if $\{\pi(k)\}$ is not a constant sequence, then the balls of $\|\cdot\|_{\pi(k)}$ semi-norm would be time-varying.

Using this interpretation for a random adapted chain $\{W(k)\}$, Lemma 4.5 asserts that the conditional expectation of the *random (semi)-distance* of the *random point* $x(k)$ from the consensus line is non-increasing function of k.

Another interesting geometric aspect of the averaging dynamics is that the sequence $\{(\pi^T(k)x(k))e\}$ of the projection points is a martingale sequence.

Lemma 4.7 *Consider an adapted random chain $\{W(k)\}$ with an absolute probability process $\{\pi(k)\}$. Then, for any random dynamics $\{x(k)\}$ driven by $\{W(k)\}$, the scalar sequence $\{\pi^T(k)x(k)\}$ is a martingale.*

Proof For any $k \ge t_0$, we have

$$\mathsf{E}\Big[\pi^T(k+1)x(k+1) \mid \mathcal{F}_k\Big] = \mathsf{E}\Big[\pi^T(k+1)W(k)x(k) \mid \mathcal{F}_k\Big]$$
$$= \mathsf{E}\Big[\pi^T(k+1)W(k) \mid \mathcal{F}_k\Big]x(k) = \pi^T(k)x(k),$$

which holds since $x(k)$ is measurable with respect to $\mathcal{F}_k$ and $\{\pi(k)\}$ is an absolute probability process for $\{W(k)\}$. **Q.E.D.**

For a deterministic chain $\{A(k)\}$, although the sequence of semi-norms $\{\|\cdot\|_{\pi(k)}\}$ maybe time-changing, Lemma 4.7 shows that the projection sequence $\{\pi^T(k)x(k)e\}$ is time-invariant.

4.3.4 An Essential Relation

To make use of a comparison function for a given dynamics, it would be helpful to quantify the amount of decrease of the particular comparison function along trajectories of the given dynamics. Here, we derive a bound for the decrease of the quadratic comparison function along the trajectories of the averaging dynamics.

To derive such a bound for the quadratic comparison function, consider the following result in spectral graph theory.

Lemma 4.8 *Let L be a symmetric matrix, i.e. $L_{ij} = L_{ji}$ for all $i, j \in [m]$. Also let $Le = 0$. Then for any vector $x \in \mathbb{R}^m$, we have*

$$x^T L x = -\sum_{i<j} L_{ij}(x_i - x_j)^2.$$

Proof ([10], p. 2.2) For any $i, j \in [m]$, let $B^{(i,j)}$ be the $m \times m$ matrix with $B_{ii}^{(i,j)} = B_{jj}^{(i,j)} = 1$, $B_{ij}^{(i,j)} = B_{ji}^{(i,j)} = -1$ and all the other entries equal to zero, or in other words $B^{(i,j)} = (e_i - e_j)(e_i - e_j)^T$. Note that $x^T B^{(i,j)} x = (x_i - x_j)^2$. Since L is symmetric and its rows sum up to zero, we have $L = \sum_{i<j} L_{ij} B^{(i,j)}$. This can be verified by observing that all the entries of the matrices on the left-hand side and the right-hand side of this equation are equivalent. Thus, we have:

$$x^T L x = x^T \left(\sum_{i<j} L_{ij} B^{(i,j)}\right) x = -\sum_{i<j} L_{ij}(x_i - x_j)^2.$$

Q.E.D.

Based on the preceding result, we can provide a lower bound on the rate of convergence of the quadratic comparison function along any trajectory of the random dynamics.

Theorem 4.4 *Let $\{W(k)\}$ be a random chain adapted to the filtration $\{\mathcal{F}_k\}$ and let $\{\pi(k)\}$ be an absolute probability process for $\{W(k)\}$. Then, for any trajectory $\{x(k)\}$ under $\{W(k)\}$, we have almost surely for all $k \geq 0$,*

$$\mathsf{E}\big[V_\pi(x(k+1), k+1) \mid \mathcal{F}_k\big] \leq V_\pi(x(k), k) - \sum_{i<j} H_{ij}(k)(x_i(k) - x_j(k))^2, \quad (4.8)$$

where $H(k) = \mathsf{E}\big[W(k)^T \mathrm{diag}(\pi(k+1))W(k) \mid \mathcal{F}_k\big]$. Furthermore, if $\pi^T(k+1) W(k) = \pi^T(k)$ holds almost surely for all $k \geq 0$, then the above inequality holds as equality.

Proof Note that by Lemma 4.7, the sequence $\{\pi^T(k)x(k)\}$ is a martingale sequence and since $f(s) = s^2$ is a convex function, by Lemma A.4, it follows that $\{-(\pi^T(k)x(k))^2\}$ is a super-martingale. Also, we have

$$V_\pi(x(k), k) = \sum_{i=1}^{m} \pi_i(k)x_i^2(k) - (\pi^T(k)x(k))^2 = x^T(k)\mathrm{diag}(\pi(k))x(k) - (\pi^T(k)x(k))^2.$$

$$(4.9)$$

Thus, if we let $\Delta(x(k), k) = V_\pi(x(k), k) - V_\pi(x(k+1), k+1)$, then using $x(k+1) = W(k)x(k)$, we have

$$
\begin{aligned}
\Delta(x(k), k) &= x^T(k)\mathrm{diag}(\pi(k))x(k) - (\pi^T(k)x(k))^2 \\
&\quad - \left\{ x^T(k+1)\mathrm{diag}(\pi(k+1))x(k+1) - (\pi^T(k+1)x(k+1))^2 \right\} \\
&= x^T(k)\left[\mathrm{diag}(\pi(k)) - W^T(k)\mathrm{diag}(\pi(k+1))W(k) \right]x(k) \\
&\quad + \left\{ (\pi^T(k+1)x(k+1))^2 - (\pi^T(k)x(k))^2 \right\} \\
&= x^T(k)L(k)x(k) + \left\{ (\pi^T(k+1)x(k+1))^2 - (\pi^T(k)x(k))^2 \right\},
\end{aligned}
$$

where $L(k) = \mathrm{diag}(\pi(k)) - W^T(k)\mathrm{diag}(\pi(k+1))W(k)$.

Since $\{-(\pi^T(k)x(k))^2\}$ is a super-martingale, by taking the conditional expectation on both sides of the above equality, we have:

$$\mathsf{E}\left[\Delta(x(k), k) \mid \mathcal{F}_k\right] \geq \mathsf{E}\left[x^T(k)L(k)x(k) \mid \mathcal{F}_k\right] = x^T(k)H(k)x(k) \qquad \text{a.s.}$$

$$(4.10)$$

where $H(k) = \mathsf{E}\left[L(k) \mid \mathcal{F}_k\right]$ and the last equality holds since $x(k)$ is measurable with respect to $\mathcal{F}_k$. Note that if $\pi^T(k+1)x(k+1) = \pi^T(k)x(k)$ almost surely, then the relation (4.10) holds as an equality. Since, this relation is the only place where we encounter an inequality, all the upcoming inequalities hold as equality if we almost surely have $\pi^T(k+1)x(k+1) = \pi^T(k)x(k)$ (as in the case of deterministic dynamics).

Furthermore, we almost surely have:

$$
\begin{aligned}
H(k)e &= \mathsf{E}\left[\mathrm{diag}(\pi(k))e - W^T(k)\mathrm{diag}(\pi(k+1))W(k)e \mid \mathcal{F}_k\right] \\
&= \pi(k) - \mathsf{E}\left[W^T(k)\pi(k+1) \mid \mathcal{F}_k\right] = 0,
\end{aligned}
$$

which is true since $W(k)$ is stochastic almost surely and $\{\pi(k)\}$ is an absolute probability process for $\{W(k)\}$. Thus, although $H(k)$ is a random matrix, it is symmetric and furthermore $H(k)e = 0$ almost surely. Thus by Lemma 4.8, we almost surely have $x^T(k)H(k)x(k) = -\sum_{i<j} H_{ij}(k)(x_i(k) - x_j(k))^2$. Using this relation in the inequality (4.10), we conclude that almost surely:

$$\mathsf{E}\left[V_\pi(x(k+1), k+1) \mid \mathcal{F}_k\right] \leq V_\pi(x(k), k) - \sum_{i<j} H_{ij}(k)(x_i(k) - x_j(k))^2.$$

Q.E.D.

Theorem 4.4 is a generalization of Lemma 4 in [6] and its generalization provided in Theorem 5 in [11].

Furthermore, the relation in Theorem 4.4 is the central relation which serves as a basis for several results in the forthcoming sections.

One of the important implications of Theorem 4.4 is the following result.

Corollary 4.3 *Let $\{\pi(k)\}$ be an absolute probability process for an adapted random chain $\{W(k)\}$. Then, for any random dynamics $\{x(k)\}$ driven by $\{W(k)\}$, we have for any $t_0 \geq 0$,*

$$\mathsf{E}\left[\sum_{k=t_0}^{\infty}\sum_{i<j} L_{ij}(k)\left(x_i(k) - x_j(k)\right)^2\right] \leq \mathsf{E}\left[V_\pi(x(t_0), t_0)\right] < \infty,$$

where $L(k) = W^T(k)\mathrm{diag}(\pi(k+1))W(k)$.

Proof By taking expectation on both sides of the relation (4.8), we have:

$$\mathsf{E}\left[V_\pi(x(k+1), k+1)\right] \leq \mathsf{E}\left[V_\pi(x(k), k)\right] - \mathsf{E}\left[\sum_{i<j}\mathsf{E}[L_{ij}(k) \mid \mathcal{F}_k](x_i(k) - x_j(k))^2\right].$$

$$(4.11)$$

Note that $\mathsf{E}[L_{ij}(k) \mid \mathcal{F}_k](x_i(k) - x_j(k))^2 = \mathsf{E}[L_{ij}(k)(x_i(k) - x_j(k))^2 \mid \mathcal{F}_k]$ holds since $x(k)$ is measurable with respect to $\mathcal{F}_k$. Also, by Lemma A.3, we have

$$\mathsf{E}\left[\mathsf{E}\left[\sum_{i<j} L_{ij}(k)(x_i(k) - x_j(k))^2 \mid \mathcal{F}_k\right]\right] = \mathsf{E}\left[\sum_{i<j} L_{ij}(k)(x_i(k) - x_j(k))^2\right].$$

Using this relation in Eq. (4.11), we have

$$\mathsf{E}\left[V_\pi(x(k+1), k+1)\right] \leq \mathsf{E}\left[V_\pi(x(k), k)\right] - \mathsf{E}\left[\sum_{i<j} L_{ij}(k)(x_i(k) - x_j(k))^2\right],$$

and hence, $\sum_{k=t_0}^{\infty}\mathsf{E}\left[\sum_{i<j} L_{ij}(k)(x_i(k) - x_j(k))^2\right] \leq \mathsf{E}[V_\pi(x(t_0), t_0)]$ for any $t_0 \geq 0$. Therefore, by Corollary A.2, we have

$$\mathsf{E}\left[\sum_{k=t_0}^{\infty}\sum_{i<j} L_{ij}(k)(x_i(k) - x_j(k))^2\right] \leq \mathsf{E}\left[V_\pi(x(t_0), t_0)\right].$$

Q.E.D.

4.4 Class $\mathcal{P}^*$

Now, we are ready to introduce a special class of random chains which we refer to as the *class $\mathcal{P}^*$*. We prove one of the central results of this thesis, i.e. any chain in class $\mathcal{P}^*$ with weak feedback property is infinite flow stable.

Definition 4.5 (*Class* $\mathcal{P}^*$) We let the class $\mathcal{P}^*$ be the class of random adapted chains that have an absolute probability process $\{\pi(k)\}$ such that $\pi(k) \geq p^*$ almost surely for some scalar $p^* > 0$ and all $k \geq 0$.

It may appear that the definition of class $\mathcal{P}^*$ is rather a restrictive requirement. In the following section, we show that in fact class $\mathcal{P}^*$ contains a broad class of deterministic and random chains.

To show that any chain in class $\mathcal{P}^*$ with weak feedback property is infinite flow stable, we prove a sequence of intermediate results. The first result gives a lower bound for the amount of the flow that is needed to ensure a decrease in the distance between the opinion of agents in a set S and its complement.

Lemma 4.9 *Let $\{A(k)\}$ be a deterministic sequence, and let $\{z(k)\}$ be generated by $z(k + 1) = A(k)z(k)$ for all $k \geq 0$ with an initial state $z(0) \in \mathbb{R}^m$. Then, for any nontrivial subset $S \subset [m]$ and $k \geq 0$, we have*

$$\max_{i \in S} z_i(k + 1) \leq \max_{s \in S} z_s(0) + d(z(0)) \sum_{t=0}^{k} A_S(t),$$

$$\min_{j \in \bar{S}} z_j(k + 1) \geq \min_{r \in \bar{S}} z_r(0) - d(z(0)) \sum_{t=0}^{k} A_S(t),$$

where $d(y) = \max_{\ell \in [m]} y_\ell - \min_{r \in [m]} y_r$ for $y \in \mathbb{R}^m$.

Proof Let $S \subset [m]$ be an arbitrary nontrivial set and let $k \geq 0$ be arbitrary. Let $z_{\min}(k) = \min_{r \in [m]} z_r$ and $z_{\max}(k) = \max_{s \in [m]} z_s(k)$. Since $z_i(k + 1) = \sum_{\ell=1}^{m} A_{i\ell}(k)z_\ell(k)$ and $A(k)$ is stochastic, we have $z_i(k) \in [z_{\min}(0), z_{\max}(0)]$ for all $i \in [m]$ and all k. Then, we obtain for $i \in S$,

$$z_i(k + 1) = \sum_{\ell \in S} A_{i\ell}(k)z_\ell(k) + \sum_{\ell \in \bar{S}} A_{i\ell}(k)z_\ell(k) \leq \sum_{\ell \in S} A_{i\ell}(k) \left(\max_{s \in S} z_s(k) \right) + z_{\max}(0) \sum_{\ell \in \bar{S}} A_{i\ell}(k),$$

where the inequality follows by $A_{i\ell}(k) \geq 0$. By the stochasticity of $A(k)$, we also obtain

$$z_i(k + 1) \leq \left(1 - \sum_{\ell \in \bar{S}} A_{i\ell}(k) \right) \max_{s \in S} z_s(k) + z_{\max}(0) \sum_{\ell \in \bar{S}} A_{i\ell}(k)$$

$$= \max_{s \in S} z_s(k) + \left(z_{\max}(0) - \max_{s \in S} z_s(k) \right) \sum_{\ell \in \bar{S}} A_{i\ell}(k).$$

By the definition of $A_S(k)$, we have $0 \leq \sum_{\ell \in \bar{S}} A_{i\ell}(k) \leq A_S(k)$. Since $z_{\max}(0) - \max_{s \in S} z_s(k) \geq 0$, it follows

$$z_i(k+1) \leq \max_{s \in S} z_s(k) + (z_{\max}(0) - \max_{s \in S} z_s(k))A_S(k) \leq \max_{s \in S} z_s(k) + d(z(0))A_S(k),$$

where the last inequality holds since $z_{\max}(0) - \max_{s \in S} z_s(k) \leq z_{\max}(0) - z_{\min}(0) = d(z(0))$. By taking the maximum over all $i \in S$ in the preceding relation, we obtain $\max_{i \in S} z_i(k+1) \leq \max_{s \in S} z_s(k) + d(z(0))A_S(k)$ and recursively, we get $\max_{i \in S} z_i(k+1) \leq \max_{s \in S} z_s(0) + d(z(0)) \sum_{t=0}^{k} A_S(t)$.

The relation for $\min_{j \in \bar{S}} z(k+1)$ follows from the preceding relation by considering $\{z(k)\}$ generated with the starting point $-z(0)$. **Q.E.D.**

Based on Lemma 4.9, we can prove the following result.

Lemma 4.10 *Let $\{A(k)\}$ be a deterministic chain with infinite flow graph $G^\infty = ([m], \mathcal{E}^\infty)$. Let $(t_0, x(t_0)) \in \mathbb{Z}^+ \times \mathbb{R}^m$ be an initial condition for the dynamics driven by $\{A(k)\}$. Then, if $\lim_{k \to \infty} (x_{i_0}(k) - x_{j_0}(k)) \neq 0$ for some i_0, j_0 belonging to the same connected component of G^∞, then we have*

$$\sum_{k=t_0}^{\infty} \sum_{i<j} [(A_{ij}(k) + A_{ji}(k))(x_i(k) - x_j(k))^2] = \infty.$$

Proof Let i_0, j_0 be in a same connected component of G^∞ with $\limsup_{k \to \infty} (x_{i_0}(k) - x_{j_0}(k)) = \alpha > 0$. Without loss of generality we may assume that $x(t_0) \in [-1, 1]^m$, otherwise we can consider the dynamics started at $y(t_0) = \frac{1}{\|x(t_0)\|_\infty} x(t_0)$. Let S be the vertices of the connected component of G^∞ containing i_0, j_0 and without loss of generality assume $S = \{1, 2, \ldots, q\}$. Then by the definition of infinite flow graph, there exists a large enough $K \geq 0$ such that $\sum_{k=K}^{\infty} A_S(k) \leq \frac{\alpha}{32q}$. Also, since $\limsup_{k \to \infty} (x_{i_0}(k) - x_{j_0}(k)) = \alpha > 0$, there exists a time instance $t_1 \geq K$ such that $x_{i_0}(t_1) - x_{j_0}(t_1) \geq \frac{\alpha}{2}$.

Let $\pi : [q] \to [q]$ be a permutation such that $x_{\pi(1)}(t_1) \geq x_{\pi(2)}(t_1) \geq \cdots \geq x_{\pi(q)}(t_1)$, i.e. π is an ordering of $\{x_i(t_1) \mid i \in [q]\}$. Then, since $x_{i_0}(t_1) - x_{j_0}(t_1) \geq \frac{\alpha}{2}$, it follows that $x_{\pi(1)}(t_1) - x_{\pi(q)}(t_1) \geq \frac{\alpha}{2}$ and, hence, there exists $\ell \in [q]$ such that $x_{\pi(\ell)}(t_1) - x_{\pi(\ell+1)}(t_1) \geq \frac{\alpha}{2q}$. Let

$$T_1 = \arg \min_{t > t_1} \sum_{k=t_1}^{t} \sum_{\substack{i,j \in [q] \\ i \leq \ell, \ell+1 \leq j}} (A_{\pi(i)\pi(j)}(k) + A_{\pi(j)\pi(i)}(k)) \geq \frac{\alpha}{32q}.$$

Note that since S is a connected component of G^∞, we should have $T_1 < \infty$, otherwise, S can be decomposed into two disconnected components $R = \{\pi(1), \ldots, \pi(l)\}$ and $S \setminus R = \{\pi(l+1), \ldots, \pi(q)\}$.

Now, if we let $R = \{\pi(1), \ldots, \pi(l)\}$, for any $k \in [t_1, T_1]$, we have:

$$\sum_{k=t_1}^{T_1-1} A_R(k) = \sum_{k=t_1}^{T_1-1} \left(\sum_{\substack{i,j \in [q] \\ i \le \ell, \ell+1 \le j}} (A_{\pi(i)\pi(j)}(k) + A_{\pi(j)\pi(i)}(k)) \right.$$

$$\left. + \sum_{i \le \ell, j \in \bar{S}} A_{\pi(i)j}(k) + \sum_{i \in \bar{S}, j \le l} A_{i\pi(j)}(k) \right)$$

$$\le \sum_{k=t_1}^{T_1-1} \sum_{\substack{i,j \in [q] \\ i \le \ell, \ell+1 \le j}} (A_{\pi(i)\pi(j)}(k) + A_{\pi(j)\pi(i)}(k)) + \sum_{k=K}^{\infty} A_S(k) \le \frac{\alpha}{16q},$$

which follows by the definition of T_1 and the choice $t_1 \ge K$. Thus by Lemma 4.9, it follows that for $k \in [t_1, T_1]$, we have $\max_{i \in R} x_i(k) \le \max_{i \in R} x_i(t_1) + 2\frac{\alpha}{16q}$. Similarly, we have $\min_{i \in S \setminus R} x_i(k) \ge \min_{i \in S \setminus R} x_i(t_1) - 2\frac{\alpha}{16q}$.

Thus, for any $i, j \in [q]$ with $i \le l$ and $j \ge l+1$, and for any $k \in [t_1, T_1]$, we have

$$x_{\pi(i)}(k) - x_{\pi(j)}(k) \ge 2(2\frac{\alpha}{16q}) = \frac{\alpha}{4q}.$$

Therefore,

$$\sum_{k=t_1}^{T_1} \sum_{\substack{i,j \in [q] \\ i \le \ell, \ell+1 \le j}} (A_{\pi(i)\pi(j)}(k) + A_{\pi(j)\pi(i)}(k))(x_{\pi(i)}(k) - x_{\pi(j)}(k))^2$$

$$\ge \left(\frac{\alpha}{4q}\right)^2 \sum_{k=t_0}^{T_1} \sum_{\substack{i,j \in [q] \\ i \le l, j \ge l+1}} (A_{\pi(i)\pi(j)}(k) + A_{\pi(j)\pi(i)}(k)) \ge (\frac{\alpha}{4q})^2 \frac{\alpha}{32q} = \beta > 0.$$

Thus, it follows that:

$$\sum_{k=t_1}^{T_1} \sum_{i<j} (A_{ij}(k) + A_{ji}(k))(x_i(k) - x_j(k))^2$$

$$\ge \sum_{k=t_1}^{T_1} \sum_{\substack{i,j \in [q] \\ i \le \ell, \ell+1 \le j}} (A_{\pi(i)\pi(j)}(k) + A_{\pi(j)\pi(i)}(k))(x_{\pi(i)}(k) - x_{\pi(j)}(k))^2 \ge \beta.$$

But since $\limsup_{k \to \infty} (x_{i_0}(k) - x_{j_0}(k)) = \alpha > 0$, there exists a time $t_2 > T_1$ such that $x_{i_0}(t_2) - x_{j_0}(t_2) \ge \frac{\alpha}{2}$. Then, using the above argument, there exists $T_2 > t_2$ such that $\sum_{k=t_2}^{T_2} \sum_{i<j} (A_{ij}(k) + A_{ji}(k))(x_i(k) - x_j(k))^2 \ge \beta$. Thus, using the induction, we can find time instances

$$\cdots > T_{\xi+1} > t_{\xi+1} > T_\xi > t_\xi > T_{\xi-1} > t_{\xi-1} > \cdots > T_1 > t_1,$$

such that $\sum_{k=t_\xi}^{T_\xi} \sum_{i<j} (A_{ij}(k) + A_{ji}(k))(x_i(k) - x_j(k))^2 \geq \beta$ for any $\xi \geq 1$. Since, the intervals $[t_\xi, T_\xi]$ are non-overlapping subintervals of $[t_0, \infty)$, we conclude that

$$\sum_{k=t_0}^{\infty} \sum_{i<j} (A_{ij}(k) + A_{ji}(k))(x_i(k) - x_j(k))^2 = \infty.$$

Q.E.D.

Now, we can prove the main result of this section.

Theorem 4.5 *Let $\{W(k)\} \in \mathcal{P}^*$ be an adapted random chain with weak feedback property. Then, $\{W(k)\}$ is infinite flow stable.*

Proof Since $\{W(k)\} \in \mathcal{P}^*$, $\{W(k)\}$ has an absolute probability process $\{\pi(k)\} \geq p^* > 0$ almost surely. Thus it follows that

$$p^* \mathsf{E}\left[W^T(k)W(k) \mid \mathcal{F}_k\right] \leq \mathsf{E}\left[W^T(k)\mathrm{diag}(\pi(k+1))W(k) \mid \mathcal{F}_k\right] = H(k).$$

On the other hand, by the weak feedback property, we have

$$\gamma \mathsf{E}\left[W_{ij}(k) + W_{ji}(k) \mid \mathcal{F}_k\right] \leq \mathsf{E}\left[W^{iT}(k)W^j(k) \mid \mathcal{F}_k\right],$$

for some $\gamma \in (0, 1]$ and for all distinct $i, j \in [m]$. Thus, we have $p^*\gamma \mathsf{E}\left[W_{ij}(k) + W_{ji}(k) \mid \mathcal{F}_k\right] \leq H_{ij}(k)$. Therefore, for the random dynamics $\{x(k)\}$ driven by $\{W(k)\}$ started at arbitrary $t_0 \geq 0$ and $x(t_0) \in \mathbb{R}^m$, by using Theorem 4.4, we have

$$\mathsf{E}[V_\pi(x(k+1), k+1) \mid \mathcal{F}_k] \leq V_\pi(x(k), k) - p^*\gamma \sum_{i<j} \mathsf{E}[W_{ij}(k) + W_{ji}(k) \mid \mathcal{F}_k](x_i(k) - x_j(k))^2.$$

By taking expectation on both sides, we obtain:

$$\mathsf{E}[\mathsf{E}[V_\pi(x(k+1), k+1) \mid \mathcal{F}_k]] \leq \mathsf{E}[V_\pi(x(k), k)]$$
$$- p^*\gamma \mathsf{E}\left[\sum_{i<j} \mathsf{E}[W_{ij}(k) + W_{ji}(k) \mid \mathcal{F}_k](x_i(k) - x_j(k))^2\right]. \qquad (4.12)$$

Since $x(k)$ is measurable with respect to $\mathcal{F}_k$, it follows

$$\mathsf{E}[W_{ij}(k) + W_{ji}(k) \mid \mathcal{F}_k](x_i(k) - x_j(k))^2 = \mathsf{E}\left[(W_{ij}(k) + W_{ji}(k))(x_i(k) - x_j(k))^2 \mid \mathcal{F}_k\right].$$

Therefore, by Lemma A.3, we obtain

$$\mathsf{E}\left[\mathsf{E}\left[\sum_{i<j}(W_{ij}(k) + W_{ji}(k))(x_i(k) - x_j(k))^2 \mid \mathcal{F}_k\right]\right]$$

$$= \mathsf{E}\left[\sum_{i<j}(W_{ij}(k) + W_{ji}(k))(x_i(k) - x_j(k))^2\right].$$

Using this relation in Eq. (4.12) and using Lemma A.3, we have

$$\mathsf{E}\left[V_\pi(x(k+1), k+1)\right] \le \mathsf{E}\left[V_\pi(x(k), k)\right] - p^*\gamma\mathsf{E}\left[\sum_{i<j}(W_{ij}(k) + W_{ji}(k))(x_i(k) - x_j(k))^2\right],$$

and hence, $p^*\gamma \sum_{k=t_0}^{\infty} \mathsf{E}\left[\sum_{i<j}(W_{ij}(k) + W_{ji}(k))(x_i(k) - x_j(k))^2\right] \le \mathsf{E}\left[V_\pi(x(t_0), t_0)\right]$.

By Corollary A.2, we have

$$\mathsf{E}\left[\sum_{k=t_0}^{\infty}\sum_{i<j}(W_{ij}(k) + W_{ji}(k))(x_i(k) - x_j(k))^2\right] \le \mathsf{E}[V_\pi(x(t_0), t_0)]$$

implying almost surely,

$$\sum_{k=t_0}^{\infty}\sum_{i<j}(W_{ij}(k) + W_{ji}(k))(x_i(k) - x_j(k))^2 < \infty.$$

Therefore, by Lemma 4.10, we conclude that $\lim_{k\to\infty}\left(x_i(k, \omega) - x_j(k, \omega)\right) = 0$ for any i, j belonging to the same connected component of $G^\infty(\omega)$, for almost all $\omega \in \Omega$. **Q.E.D.**

Theorem 4.5 not only shows that the dynamics (3.1) is asymptotically stable almost surely for chains in $\mathcal{P}^*$ with weak feedback property, but also characterizes the equilibrium points of such dynamics.

Using the alternative characterization of the infinite flow stability in Lemma 4.1, we have the following result.

Corollary 4.4 *Let $\{W(k)\}$ be an adapted random chain in $\mathcal{P}^*$ that has weak feedback property. Then, $\lim_{k\to\infty} W(k : t_0)$ exists almost surely for all $t_0 \ge 0$.*

In the following section, we characterize a broad class of chains that belong to the class $\mathcal{P}^*$. Before characterizing this subclass, let us discuss a straightforward generalization of Theorem 4.5 to any (not necessarily adapted) random chain but assuming a stronger condition.

Corollary 4.5 *Let $\{W(k)\}$ be a random chain such that almost all of its sample paths are in $\mathcal{P}^*$ and almost all sample paths have weak feedback property. Then, $\{W(k)\}$ is an infinite flow stable chain.*

Proof Note that if a sample path $\{W(k, \omega)\}$ is in $\mathcal{P}^*$ and has weak feedback property, then Theorem 4.5 applies to that particular sample path $\{W(k, \omega)\}$. Thus, if almost all the sample paths are satisfying these properties, we can conclude that the random chain is infinite flow stable. **Q.E.D.**

4.5 Balanced Chains

In this section, we characterize a subclass of $\mathcal{P}^*$ chains and independent random chains, namely the class of *balanced chains* with feedback property. We first show that this class includes many of the chains that have been studied in the existing literature. Then we show that any balanced chain with feedback property has a uniformly bounded absolute probability process, or in other words, it belongs to the class $\mathcal{P}^*$.

We start our development by considering deterministic chains and later, we discuss the random counterparts. For this, let $\{A(k)\}$ be a deterministic chain of stochastic matrices and let $A_{S\bar{S}}(k) = \sum_{i \in S, j \in \bar{S}} A_{ij}(k)$ for a non-empty $S \subset [m]$. We define balanced chains as follows.

Definition 4.6 A chain $\{A(k)\}$ is balanced if there exists a scalar $\alpha > 0$, such that

$$A_{S\bar{S}}(k) \geq \alpha A_{\bar{S}S}(k) \quad \text{for any non-trivial } S \subset [m] \text{ and } k \geq 0. \tag{4.13}$$

We refer to α as a balancedness coefficient.

Note that in Definition 4.6, the scalar α is time-independent. Furthermore, due to the inter-changeability of any non-trivial subset S with its complement $\bar{S}$, for a balanced chain $\{A(k)\}$ we have

$$A_{S\bar{S}}(k) \geq \alpha A_{\bar{S}S}(k) \geq \alpha^2 A_{S\bar{S}}(k),$$

implying $\alpha \leq 1$.

Let us first discuss the graph theoretic interpretation of the balancedness property. For this, note that every $m \times m$ matrix A can be viewed as a directed weighted graph $G = ([m], [m] \times [m], A)$, where the edge (i, j) has the weight A_{ij}. For example, in Fig. 4.3, the directed graph representation of the matrix

$$\begin{bmatrix} \frac{1}{2} & 0 & \frac{1}{2} \\ \frac{1}{2} & 0 & \frac{1}{2} \\ 0 & 0 & 1 \end{bmatrix} \tag{4.14}$$

is depicted.

In the graph representation of a matrix A, given a non-trivial subset $S \subset [m]$, the set $S \times \bar{S}$ corresponds to the set of edges going out from the subset S. With this in mind, $A_{S\bar{S}}$ is the sum of the weights going out of S, i.e. the *flow* going out from S.

Fig. 4.3 The representation of the matrix defined in Eq. (4.14) with a directed weighted graph

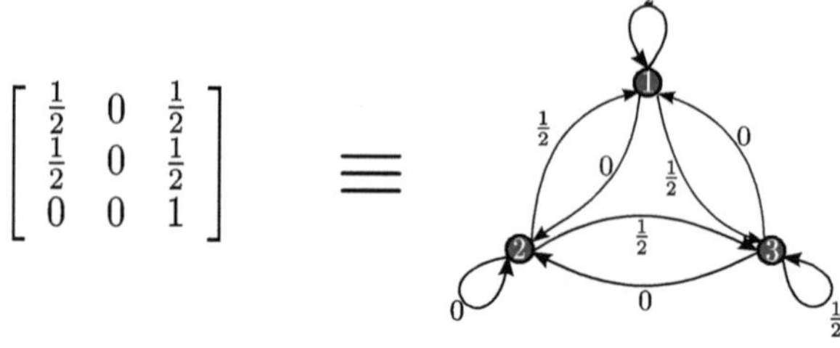

On the other hand, $A_{\bar{S}S}$ is the flow going out from $\bar{S}$ or in other words the flow entering the set S. Balancedness property requires that the ratio between the flow going out from each subset of vertices and the flow entering the subset does not vanish across the time.

Note that we can extend the notion of balanced chains to the chains that are balanced from some time instance $t_0 \geq 0$ onward, i.e. the chains such that $A_{S\bar{S}}(k) \geq \alpha A_{\bar{S}S}(k)$ for some $\alpha > 0$, any non-trivial $S \subset [m]$, and any $k \geq t_0$. In this case, all the subsequent analysis can be applied to the sub-chain $\{A(k)\}_{k \geq t_0}$.

Before continuing our analysis of the balanced chains, let us discuss some of the well-known subclasses of these chains:

Balanced Bidirectional Chains

We say that $\{A(k)\}$ is a balanced bidirectional chain if there exists some $\alpha > 0$ such that $A_{ij}(k) \geq \alpha A_{ji}(k)$ for any $k \geq 0$ and $i, j \in [m]$. These chains are in fact balanced, since, for any $S \subset [m]$, we have:

$$A_{S\bar{S}}(k) = \sum_{i \in S, j \in \bar{S}} A_{ij}(k) \geq \sum_{i \in S, j \in \bar{S}} \alpha A_{ji}(k) = \alpha A_{\bar{S}S}(k).$$

Examples of such chains are bounded bidirectional chains, that are chains such that $A_{ij}(k) > 0$ implies $A_{ji}(k) > 0$ and also positive entries are uniformly bounded by some $\gamma > 0$. In other words, $A_{ij}(k) > 0$ implies $A_{ij}(k) \geq \gamma$ for any $i, j \in [m]$. In this case, for $A_{ij}(k) > 0$, we have $A_{ij}(k) \geq \gamma \geq \gamma A_{ji}(k)$ and for $A_{ij}(k) = 0$, we have $A_{ji}(k) = 0$ and hence, in either of the cases $A_{ij}(k) \geq \gamma A_{ji}(k)$. Therefore, bounded bidirectional chains are examples of balanced bidirectional chains. Such chains have been considered in [12–14]. The reverse implication is not necessarily true since positive entries of $\{A(k)\}$ can go to zero but yet maintain the bounded bidirectional property.

B-connected Chains

Let $\{A(k)\}$ be a B-connected chain (see Definition 2.4). Such a chain may not be balanced originally, however, if we instead consider the chain $\{\tilde{A}(k)\}$ defined by $\tilde{A}(k) = A((k + 1)B : kB)$, then $\{\tilde{A}(k)\}$ is balanced. The reason is that since $G(k) = ([m], \mathcal{E}(Bk) \cup \mathcal{E}(Bk + 1) \cup \cdots \cup \mathcal{E}(B(k + 1) - 1))$ is strongly connected, for any $S \subset [m]$, we have an edge $(i, j) \in S \times \bar{S}$ with $\tilde{A}_{ij}(k) \geq \gamma^m$. Therefore,

$$\tilde{A}_{S\bar{S}}(k) \geq \gamma^m \geq \frac{\gamma^m}{m} \tilde{A}_{\bar{S}S}(k).$$

Hence, $\{\tilde{A}(k)\}$ is a balanced chain.

Chains with Common Steady State $\pi > 0$

This ensemble consists of chains $\{A(k)\}$ with $\pi^T A(k) = \pi^T$ for some stochastic vector $\pi > 0$ and all $k \geq 0$, which are generalizations of doubly stochastic chains, where we have $\pi = \frac{1}{m}e$. Doubly stochastic chains and chains with common steady state $\pi > 0$ are the subject of the studies in [6, 11, 15].
To show that a chain with a common steady state $\pi > 0$ is an example of a balanced chain, let us prove the following lemma.

Lemma 4.11 *Let A be a stochastic matrix and $\pi > 0$ be a stochastic left-eigenvector of A corresponding to the unit eigenvalue, i.e., $\pi^T A = \pi^T$. Then, $A_{S\bar{S}} \geq \frac{\pi_{\min}}{\pi_{\max}} A_{\bar{S}S}$ for any non-trivial $S \subset [m]$, where $\pi_{\max} = \max_{i \in [m]} \pi_i$ and $\pi_{\min} = \min_{i \in [m]} \pi_i$.*

Proof Let $S \subset [m]$. Since $\pi^T A = \pi^T$, we have

$$\sum_{j \in S} \pi_j = \sum_{i \in [m], j \in S} \pi_i A_{ij} = \sum_{i \in S, j \in S} \pi_i A_{ij} + \sum_{i \in \bar{S}, j \in S} \pi_i A_{ij}. \tag{4.15}$$

On the other hand, since A is a stochastic matrix, we have $\pi_i \sum_{j \in [m]} A_{ij} = \pi_i$. Therefore,

$$\sum_{i \in S} \pi_i = \sum_{i \in S} \pi_i \sum_{j \in [m]} A_{ij} = \sum_{i \in S, j \in S} \pi_i A_{ij} + \sum_{i \in S, j \in \bar{S}} \pi_i A_{ij}. \tag{4.16}$$

Comparing Eqs. (4.15) and (4.16), we have $\sum_{i \in \bar{S}, j \in S} \pi_i A_{ij} = \sum_{i \in S, j \in \bar{S}} \pi_i A_{ij}$. Therefore,

$$\pi_{\min} A_{\bar{S}S} \leq \sum_{i \in \bar{S}, j \in S} \pi_i A_{ij} = \sum_{i \in S, j \in \bar{S}} \pi_i A_{ij} \leq \pi_{\max} A_{S\bar{S}}.$$

Hence, we have $A_{S\bar{S}} \geq \frac{\pi_{\min}}{\pi_{\max}} A_{\bar{S}S}$ for any non-trivial $S \subset [m]$. **Q.E.D.**

The above lemma shows that chains with common steady state $\pi > 0$ are examples of balanced chains with balancedness coefficient $\alpha = \frac{\pi_{\min}}{\pi_{\max}}$. In fact, the lemma yields a much more general result, as provided below.

Theorem 4.6 *Let $\{A(k)\}$ be a stochastic chain with a sequence $\{\pi(k)\}$ of unit left eigenvectors, i.e., $\pi^T(k)A(k) = \pi^T(k)$. If $\{\pi(k)\} \geq p^*$ for some scalar $p^* > 0$, then $\{A(k)\}$ is a balanced chain with a balancedness coefficient $\alpha = \frac{p^*}{1-(m-1)p^*}$.*

Proof As proven in Lemma 4.11, we have

$$A_{S\bar{S}}(k) \geq \frac{\pi_{\min}(k)}{\pi_{\max}(k)} A_{\bar{S}S}(k),$$

for any non-trivial $S \subset [m]$ and $k \geq 0$, which follows by $\pi^T(k)A(k) = \pi^T(k)$. But if $\{\pi(k)\} \geq p^* > 0$, then $\pi_{\min}(k) \geq p^*$ for any $k \geq 0$, and since $\pi(k)$ is a stochastic vector, we have $\pi_{\max}(k) \leq 1 - (m-1)\pi_{\min}(k) \leq 1 - (m-1)p^*$. Therefore,

$$A_{S\bar{S}}(k) \geq \frac{\pi_{\min}(k)}{\pi_{\max}(k)} A_{\bar{S}S}(k) \geq \frac{p^*}{1 - (m-1)p^*} A_{\bar{S}S}(k),$$

for any non-trivial $S \subset [m]$. Thus, $\{A(k)\}$ is balanced with a coefficient $\alpha = \frac{p^*}{1-(m-1)p^*}$. **Q.E.D**.

Note that if we define a balanced chain to be a chain that is balanced from some time $t_0 \geq 0$, then in Theorem 4.6, it suffice to have $\liminf_{k\to\infty} \pi_i(k) \geq p^* > 0$ for all $i \in [m]$.

Theorem 4.6 not only characterizes a class of balanced chains, but also provides a computational way to verify balancedness. Thus, instead of verifying Definition 4.6 for every $S \subset [m]$, it suffice to find a sequence $\{\pi(k)\}$ of unit left eigenvectors of the chain $\{A(k)\}$ such that the entries of such a sequence do not vanish as time goes to infinity.

Now, let us generalize the definition of balanced chains to a more general setting of independent chains.

Definition 4.7 Let $\{W(k)\}$ be an independent random chain. We say $\{W(k)\}$ is balanced chain if the expected chain $\{\bar{W}(k)\}$ is balanced, i.e., $\bar{W}_{S\bar{S}}(k) \geq \alpha \bar{W}_{\bar{S}S}(k)$ for any $S \subset [m]$, any $k \geq 0$, and some $\alpha > 0$.

Immediate examples of such chains are deterministic balanced chains.

4.5.1 Absolute Probability Sequence for Balanced Chains

In this section, we show that every balanced chain with feedback property has a uniformly bounded absolute probability sequence, i.e., our main goal is to prove that *any balanced and independent random chain with feedback property is in* $\mathcal{P}^*$.

The road map to prove this result is as follows: we first show that this result holds for deterministic chains with uniformly bounded entries. Then, using this result and geometric properties of the set of balanced chains with feedback property, we prove the statement for deterministic chains. Finally, the result follows immediately for random chains.

To prove the result for uniformly bounded deterministic chains, we employ the technique that is used to prove Proposition 4 in [12]. However, the argument given in [12] needs some extensions to fit in our more general assumption of balancedness.

For our analysis, let $\{A(k)\}$ be a deterministic chain and let

$$S_j(k) = \{\ell \in [m] \mid A_{\ell j}(k:0) > 0\}$$

be the set of indices of the positive entries in the jth column of $A(k : 0)$ for $j \in [m]$ and $k \geq 0$. Also, let $\mu_j(k)$ be the minimum value of these positive entries, i.e.,

$$\mu_j(k) = \min_{\ell \in S(k)} A_{\ell j}(k : 0) > 0.$$

Lemma 4.12 *Let $\{A(k)\}$ be a balanced chain with feedback property and with uniformly bounded positive entries, i.e., there exists a scalar $\gamma > 0$ such that $A_{ij}(k) \geq \gamma$ for $A_{ij}(k) > 0$. Then, $S_j(k) \subseteq S_j(k+1)$ and $\mu_j(k) \geq \gamma^{|S_j(k)|-1}$ for all $j \in [m]$ and $k \geq 0$.*

Proof Let $j \in [m]$ be arbitrary but fixed. By induction on k, we prove that $S_j(k) \subseteq S_j(k+1)$ for all $k \geq 0$ as well as the desired relation for $\mu_j(k)$. For $k = 0$, we have $A(0 : 0) = I$, so $S_j(0) = \{j\}$. Then, $A(1 : 0) = A(1)$ and by the feedback property of the chain $\{A(k)\}$ we have $A_{jj}(1) \geq \gamma$, implying $\{j\} = S_j(0) \subseteq S_j(1)$. Furthermore, we have $|S_j(0)| - 1 = 0$ and $\mu_j(0) = 1 \geq \gamma^0$. Hence, the claim is true for $k = 0$.

Now suppose that the claim is true for $k \geq 0$, and consider $k + 1$. Then, for any $i \in S_j(k)$, we have:

$$A_{ij}(k+1 : 0) = \sum_{\ell=1}^{m} A_{i\ell}(k) A_{\ell j}(k : 0) \geq A_{ii}(k) A_{ij}(k : 0) \geq \gamma \mu_j(k) > 0.$$

Thus, $i \in S_j(k+1)$, implying $S_j(k) \subseteq S_j(k+1)$.

To show the relation for $\mu_j(k+1)$, we consider two cases:

Case 1: $A_{S_j(k)\bar{S}_j(k)}(k) = 0$: In this case for any $i \in S_j(k)$, we have:

$$A_{ij}(k+1 : 0) = \sum_{\ell \in S_j(k)} A_{i\ell}(k) A_{\ell j}(k : 0) \geq \mu_j(k) \sum_{\ell \in S_j(k)} A_{i\ell}(k) = \mu_j(k), \quad (4.17)$$

where the inequality follows from $i \in S_j(k)$ and $A_{S_j(k)\bar{S}_j(k)}(k) = 0$, and the definition of $\mu_j(k)$. Furthermore, by the balancedness of $A(k)$ and $A_{S_j(k)\bar{S}_j(k)}(k) = 0$, it follows that $0 = A_{S_j(k)\bar{S}_j(k)}(k) \geq \alpha A_{\bar{S}_j(k)S_j(k)}(k) \geq 0$. Hence, $A_{\bar{S}_j(k)S_j(k)}(k) = 0$. Thus, for any $i \in \bar{S}_j(k)$, we have

$$A_{ij}(k+1 : 0) = \sum_{\ell=1}^{m} A_{i\ell}(k) A_{\ell j}(k : 0) = \sum_{\ell \in \bar{S}_j(k)} A_{i\ell}(k) A_{\ell j}(k : 0) = 0,$$

where the second equality follows from $A_{\ell j}(k : 0) = 0$ for all $\ell \in \bar{S}_j(k)$. Therefore, in this case we have $S_j(k+1) = S_j(k)$, which by (4.17) implies $\mu_j(k+1) \geq \mu_j(k)$. In view of $S_j(k+1) = S_j(k)$ and the inductive hypothesis, we further have

$$\mu_j(k) \geq \gamma^{|S_j(k)|-1} = \gamma^{|S_j(k+1)|-1},$$

implying $\mu_j(k+1) \geq \gamma^{|S_j(k+1)|-1}$.

Case 2: $A_{S_j(k)\bar{S}_j(k)}(k) > 0$: Since the chain is balanced, we have

$$A_{\bar{S}_j(k)S_j(k)}(k) \geq \alpha A_{S_j(k)\bar{S}_j(k)}(k) > 0,$$

implying that $A_{\bar{S}_j(k)S_j(k)}(k) > 0$. Therefore, by the uniform boundedness of the positive entries of $A(k)$, there exists $\hat{\xi} \in \bar{S}_j(k)$ and $\hat{\ell} \in S_j(k)$ such that $A_{\hat{\xi}\hat{\ell}}(k) \geq \gamma$. Hence, we have

$$A_{\hat{\xi}j}(k+1:0) \geq A_{\hat{\xi}\hat{\ell}}(k)A_{\hat{\ell}j}(k:0) \geq \gamma\mu_j(k) = \gamma^{|S_j(k)|},$$

where the equality follows by the induction hypothesis. Thus, $\hat{\xi} \in S_j(k+1)$ while $\hat{\xi} \notin S_j(k)$, which implies $|S_j(k+1)| \geq |S_j(k)| + 1$. This, together with $A_{\hat{\xi}j}(k+1:0) \geq \gamma^{|S_j(k)|}$, yields $\mu_j(k+1) \geq \gamma^{|S_j(k)|} \geq \gamma^{|S_j(k+1)|-1}$. **Q.E.D.**

Note that our proof shows that the bound for the nonnegative entries given in Proposition 4 of [12] can be reduced from γ^{m^2-m+2} to γ^{m-1}.

It can be seen that Lemma 4.12 holds for products $A(k:t_0)$ starting with any $t_0 \geq 0$ and $k \geq t_0$ (with appropriately defined $S_j(k)$ and $\mu_j(k)$). An immediate corollary of Lemma 4.12 is the following result.

Corollary 4.6 *Under the assumptions of Lemma* 4.12, *we have*

$$\frac{1}{m}e^T A(k:t_0) \geq \min(\frac{1}{m}, \gamma^{m-1})e^T,$$

for any $k \geq t_0 \geq 0$.

Proof Without loss of generality, let $t_0 = 0$. By Lemma 4.12, for any $j \in [m]$, we have $\frac{1}{m}e^T A^j(k:0) \geq \frac{1}{m}|S_j(k)|\gamma^{|S_j(k)|-1}$, where A^j denotes the jth column of A. For $\gamma \in [0,1]$, the function $t \mapsto t\gamma^{t-1}$ defined on $[1,m]$ attains its minimum at either $t = 1$ or $t = m$. Therefore, $\frac{1}{m}e^T A(k:0) \geq \min(\frac{1}{m}, \gamma^{m-1})e^T$. **Q.E.D.**

Now, we relax the assumption on the bounded entries in Corollary 4.6.

Theorem 4.7 *Let $\{A(k)\}$ be a balanced chain with feedback property. Let $\alpha, \beta > 0$ be a balancedness and feedback coefficients for $\{A(k)\}$, respectively. Then, there is a scalar $\gamma = \gamma(\alpha, \beta) \in (0,1]$ such that $\frac{1}{m}e^T A(k:0) \geq \min(\frac{1}{m}, \gamma^{m-1})e^T$ for any $k \geq 0$.*

Proof Let $\mathbf{B}_{\alpha,\beta}$ be the set of balanced matrices with the balancedness coefficient α and feedback property with coefficient $\beta > 0$, i.e.,

$$\mathbf{B}_{\alpha,\beta} := \big\{Q \in \mathbb{R}^{m\times m} \mid Q \geq 0, \, Qe = e,$$
$$Q_{S\bar{S}} \geq \alpha Q_{\bar{S}S} \text{ for all non-trivial } S \subset [m], \, Q_{ii} \geq \beta \text{ for all } i \in [m]\big\}. \tag{4.18}$$

The description in relation (4.18) shows that $\mathbf{B}_{\alpha,\beta}$ is a bounded polyhedral set in $\mathbb{R}^{m\times m}$. Let $\{Q^{(\xi)} \in \mathbf{B}_{\alpha,\beta} \mid \xi \in [n_{\alpha,\beta}]\}$ be the set of extreme points of this polyhedral set indexed by the positive integers between 1 and $n_{\alpha,\beta}$, which is the number of extreme points of $\mathbf{B}_{\alpha,\beta}$.

Since $A(k) \in \mathbf{B}_{\alpha,\beta}$ for all $k \geq 0$, we can write $A(k)$ as a convex combination of the extreme points in $\mathbf{B}_{\alpha,\beta}$, i.e., there exist coefficients $\lambda_\xi(k) \in [0, 1]$ such that

$$A(k) = \sum_{\xi=1}^{n_{\alpha,\beta}} \lambda_\xi(k) Q^{(\xi)} \quad \text{with} \sum_{\xi=1}^{n_{\alpha,\beta}} \lambda_\xi(k) = 1. \tag{4.19}$$

Now, consider the following independent random matrix process defined by:

$$W(k) = Q^{(\xi)} \quad \text{with probability } \lambda_\xi(k).$$

In view of this definition any sample path of $\{W(k)\}$ consists of extreme points of $\mathbf{B}_{\alpha,\beta}$. Thus, every sample path of $\{W(k)\}$ has a coefficient bounded by the minimum positive entry of the matrices in $\{Q^{(\xi)} \in \mathbf{B}_{\alpha,\beta} \mid \xi \in [n_{\alpha,\beta}]\}$, denoted by $\gamma = \gamma(\alpha, \beta) > 0$, where $\gamma > 0$ since $n_{\alpha,\beta}$ is finite. Therefore, by Corollary 4.6, we have $\frac{1}{m} e^T W(k : t_0) \geq \min(\frac{1}{m}, \gamma^{m-1}) e^T$ for all $k \geq t_0 \geq 0$. Furthermore, by Eq. (4.19) we have $\mathsf{E}[W(k)] = A(k)$ for all $k \geq 0$, implying

$$\frac{1}{m} e^T A(k : t_0) = \frac{1}{m} e^T \mathsf{E}\big[W(k : t_0)\big] \geq \min\left(\frac{1}{m}, \gamma^{m-1}\right) e^T,$$

which follows from $\{W(k)\}$ being independent. **Q.E.D.**

Based on the above results, we are ready to prove the main result for deterministic chains.

Theorem 4.8 *Any deterministic balanced chain with feedback property is in $\mathcal{P}^*$.*

Proof Consider a balanced chain $\{A(k)\}$ with a balancedness coefficient α and a feedback coefficient β. As in Theorem 4.1, let $\{t_r\}$ be an increasing sequence of positive integers such that $R(k) = \lim_{r\to\infty} A(t_r : k)$. As discussed in the proof of Theorem 4.2, any sequence $\{\hat{\pi}^T R(k)\}$ is an absolute probability sequence for $\{A(k)\}$, where $\hat{\pi}$ is a stochastic vector. Let $\hat{\pi} = \frac{1}{m} e$. Then, by Theorem 4.7,

$$\frac{1}{m} e^T R(k) = \frac{1}{m} \lim_{r\to\infty} e^T A(t_r : k) \geq p^* e^T,$$

with $p^* = \min(\frac{1}{m}, \gamma^{m-1}) > 0$. Thus, $\{\frac{1}{m} e^T R(k)\}$ is a uniformly bounded absolute probability sequence for $\{A(k)\}$. **Q.E.D.**

The main result of this section follows immediately from Theorem 4.8.

Theorem 4.9 *Any balanced and independent random chain with feedback property is in $\mathcal{P}^*$.*

As a result of Theorem 4.9 and Theorem 4.5, the following result holds, which is one of the central results of this thesis.

Theorem 4.10 *Let $\{W(k)\}$ be any independent random chain which is balanced and has feedback property. Then, $\{W(k)\}$ is infinite flow stable.*

Proof By Theorem 4.9, any such chain is in $\mathcal{P}^*$. Thus, Theorem 4.5 implies the result. **Q.E.D.**

As we have shown in the preceding discussions, many of the chains that are widely discussed in other literatures, are examples of balanced chains with feedback property. Thus Theorem 4.10 not only provides a unified analysis to many of the previous works in this field but also, provides conditions to which those results can be extended. As it will be shown in Chap. 5, Theorem 4.10 also shows that Lemma 1.1 holds under general assumptions for inhomogeneous and random chains of stochastic matrices.

As in Corollary 4.5, let us assert a generalization of Theorem 4.10 to an arbitrary chain.

Corollary 4.7 *Let $\{W(k)\}$ be a random chain such that its sample paths are balanced and have feedback property almost surely. Then, $\{W(k)\}$ is an infinite flow stable chain.*

References

1. F. Fagnani, S. Zampieri, Randomized consensus algorithms over large scale networks. IEEE J. Sel. Areas Commun. **26**(4), 634–649 (2008)
2. A. Tahbaz-Salehi, A. Jadbabaie, A necessary and sufficient condition for consensus over random networks. IEEE Trans. Autom. Control **53**(3), 791–795 (2008)
3. A. Tahbaz-Salehi, A. Jadbabaie, Consensus over ergodic stationary graph processes. IEEE Trans. Autom. Control **55**(1), 225–230 (2010)
4. A. Kolmogoroff, Zur Theorie der Markoffschen Ketten. Mathematische Annalen **112**(1), 155–160 (1936)
5. D. Blackwell, Finite non-homogeneous chains. Ann. Math. **46**(4), 594–599 (1945)
6. A. Nedić, A. Olshevsky, A. Ozdaglar, J. Tsitsiklis, On distributed averaging algorithms and quantization effects. IEEE Trans. Autom. Control **54**(11), 2506–2517 (2009)
7. A. Jadbabaie, J. Lin, S. Morse, Coordination of groups of mobile autonomous agents using nearest neighbor rules. IEEE Trans. Autom. Control **48**(6), 988–1001 (2003)
8. A. Olshevsky, J. Tsitsiklis, On the nonexistence of quadratic lyapunov functions for consensus algorithms. IEEE Trans. Autom. Control **53**(11), 2642–2645 (2008)
9. L. Bregman, The relaxation method of finding the common point of convex sets and its application to the solution of problems in convex programming. USSR Comput. Math. Math. Phys. **7**(3), 200–217 (1967)
10. D. Spielman, Spectral Graph Theory, The Laplacian, University Lecture Notes, 2009, available at: http://www.cs.yale.edu/homes/spielman/561/lect02-09.pdf
11. B. Touri, A. Nedić, On ergodicity, infinite flow and consensus in random models. IEEE Trans. Autom. Control **56**(7), 1593–1605 (2011)
12. J. Lorenz, A stabilization theorem for continuous opinion dynamics. Phys. A: Stat. Mech. Appl. **355**, 217–223 (2005)
13. V. Blondel, J. Hendrickx, A. Olshevsky, J. Tsitsiklis, Convergence in multiagent coordination, consensus, and flocking, in *Proceedings of IEEE CDC*, (2005)

14. M. Cao, A.S. Morse, B.D.O. Anderson, Reaching a consensus in a dynamically changing environment: A graphical approach. SIAM J. Control Optim. **47**, 575–600 (2008)
15. B. Touri, A. Nedić, On approximations and ergodicity classes in random chains, (2010) under review

Chapter 5
Implications

In this chapter, we discuss some of the implications of the results developed in Chaps. 3 and 4. We first study the implications of the developed results on the product of independent random stochastic chains in Sect. 5.1, and also develop a rate of convergence result for ergodic independent random chains. Then in Sect. 5.2, we study the implication of Theorem 4.10 in non-negative matrix theory. There, we show that Theorem 4.10 is an extension of a well-known result for homogeneous Markov chains to inhomogeneous products of stochastic matrices. In Sect. 5.3, we provide a convergence rate analysis for averaging dynamics driven by uniformly bounded chains. Then, in Sect. 5.4, we introduce link-failure models for random chains and analyze the effect of link-failure on the limiting behavior of averaging dynamics. In Sect. 5.5, we study the Hegselmann-Krause model for opinion dynamics in social networks and provide a new bound on the termination time of such dynamics. Finally, in Sect. 5.6, using the developed tools, we propose an alternative proof for the second Borel-Cantelli lemma.

5.1 Independent Random Chains

Throughout this section, we deal exclusively with an independent random chain $\{W(k)\}$ (on $\mathbb{R}^m$). For our study, let $\{\mathcal{F}_k\}$ be the natural filtration associated with $\{W(k)\}$, i.e. $\mathcal{F}_k = \sigma(W(0), \ldots, W(k-1))$ for any $k \geq 1$ and $\mathcal{F}_0 = \{\emptyset, \Omega\}$.

An interesting feature of independent random chains is that except consensus, all other events that we have discussed so far are trivial events for those chains.

Lemma 5.1 *Let $\{W(k)\}$ be an independent random chain. Then, for any $i, j \in [m]$, $i \leftrightarrow_W j$ is a tale event. Furthermore, ergodicity and infinite flow events are trivial events. Also, there exists a graph G on m vertices such that the infinite flow graph of $\{W(k)\}$ is equal to G almost surely.*

Proof Note that if for some $\omega \in \Omega$, we have $\omega \in i \leftrightarrow_W j$, then by the definition of mutual ergodicity, it follows that $\lim_{k \to \infty} \|W_i(k : t_0, \omega) - W_j(k : t_0, \omega)\| = 0$ for all $t_0 \geq 0$, where $W(k : t_0, \omega) = W(k-1, \omega) \cdots W(t_0, \omega)$. Conversely, if

B. Touri, *Product of Random Stochastic Matrices and Distributed Averaging*,
Springer Theses, DOI: 10.1007/978-3-642-28003-0_5,
© Springer-Verlag Berlin Heidelberg 2012

$\lim_{k\to\infty} \|W_i(k \,:\, t_0, \omega) - W_j(k \,:\, t_0, \omega)\| = 0$ for some $t_0 \geq 0$, then for any $t \in [t_0, 0]$ we have

$$
\begin{aligned}
\lim_{k\to\infty} \|W_i(k : t, \omega) - W_j(k : t, \omega)\| &= \lim_{k\to\infty} \|(e_i - e_j)^T W(k : t, \omega)\| \\
&= \lim_{k\to\infty} \|(e_i - e_j)^T W(k : t_0, \omega) W(t_0 : t, \omega)\| \\
&\leq \lim_{k\to\infty} \|(e_i - e_j)^T W(k : t_0, \omega)\| \, \|W(t_0 : t, \omega)\| \\
&= \|W(t_0 : t, \omega)\| \lim_{k\to\infty} \| \\
&\quad W_i(k : t_0, \omega) - W_j(k : t_0, \omega)\| = 0,
\end{aligned}
$$

where the inequality follows from the definition of the induced matrix norm. Therefore, it follows that $i \leftrightarrow_W j$ is a tale event. Since, for the ergodicity event $\mathscr{E}$ we have $\mathscr{E} = \bigcap_{i,j\in[m]} i \leftrightarrow_W j$, it follows that the ergodicity event is also a tale event. Therefore, by Kolmogorov's 0–1 law (Theorem A.5), it follows that mutual ergodicity and ergodicity events are trivial events.

To prove that the infinite flow event is a trivial event, we observe that for any non-trivial $S \subset [m]$, we have $\sum_{k=0}^{\infty} W_S(k) = \infty$ if and only if $\sum_{k=t_0}^{\infty} W_S(k) = \infty$ for any $t_0 \geq 0$, which follows from the fact that $W_S(k) \in [0, m]$ almost surely for any $k \geq 0$. Since $\{W(k)\}$ is an independent chain, the random process $\{W_S(k)\}$ is an independent chain for any $S \subset [m]$. Thus, again by application of the Kolmogorov's 0–1 law, it follows that the event $\{\omega \mid \sum_{k=0}^{\infty} W_S(k, \omega) = \infty\}$ is a trivial event. For the infinite flow event $\mathscr{F}$, we have

$$
\mathscr{F} = \bigcap_{\substack{S \subset [m] \\ S \neq \emptyset}} \{\omega \mid \sum_{k=0}^{\infty} W_S(k, \omega) = \infty\},
$$

and hence, the infinite flow event is a trivial event.

Similarly, if $G^{\infty} = ([m], \mathcal{E}^{\infty})$ is the (random) infinite flow graph of $\{W(k)\}$, then for any $i, j \in [m]$ with $i \neq j$, the event $\sum_{k=0}^{\infty} (W_{ij}(k) + W_{ji}(k)) = \infty$ is a tale event and hence, $G^{\infty-1}(G)$ is a tale event for any simple graph G. But $\{G^{\infty-1}(G) \mid G \in \mathcal{G}([m])\}$ is a partitioning of Ω and hence, $G^{\infty} = G$ almost surely for a $G \in \mathcal{G}([m])$. **Q.E.D.**

Based on Lemma 5.1 it follows that, as in the case of deterministic chains, we can simply say that an independent chain is ergodic or has an infinite flow property. Also, we can view the infinite flow graph of an independent random chain as a deterministic graph. The following lemma identifies this particular graph among the $2^{\binom{m}{2}}$ simple graphs in $\mathcal{G}([m])$.

Lemma 5.2 *The infinite flow graph G^{∞} of an independent random chain is equal to the infinite flow graph of the expected chain $\{\bar{W}(k)\}$ almost surely.*

Proof Let $\bar{G}^{\infty} = ([m], \bar{\mathcal{E}}^{\infty})$ be the infinite flow graph of the expected chain $\{\bar{W}(k)\}$. For any $i, j \in [m]$, we have $W_{ij}(k) \in [0, 1]$ almost surely. Since the sequence

$\{W_{ij}(k) + W_{ji}(k)\}$ is an independent sequence, by Corollary A.4, $\{i, j\} \in \mathcal{E}^{\infty}$ almost surely if and only if $\{i, j\} \in \bar{\mathcal{E}}^{\infty}$. Since this holds for any $i, j \in [m]$, it follows $G^{\infty} = \bar{G}^{\infty}$ almost surely. **Q.E.D.**

Recall that any absolute probability process for an independent random chain is a deterministic sequence, which is true because $\bar{W}(k) = \mathsf{E}\big[W(k) \mid \mathcal{F}_k\big]$ is a deterministic matrix for any $k \geq 0$ and any independent random chain admits such a sequence.

Based on the above observations and Theorem 4.5, we have the following result for independent random chains.

Theorem 5.1 *Let $\{W(k)\}$ be an independent random chain. Then, $\{W(k)\}$ is in $\mathcal{P}^*$ if and only if $\{\bar{W}(k)\}$ is in $\mathcal{P}^*$. Furthermore, if $\{W(k)\}$ is a chain with weak feedback property and $\{\bar{W}(k)\}$ is in $\mathcal{P}^*$, then the following properties are equivalent*

(a) $i \leftrightarrow_W j$,
(b) $i \leftrightarrow_{\bar{W}} j$,
(c) *i, j belong to a same connected component of $\bar{G}^{\infty}$,*
(d) *i, j belong to a same connected component of G^{∞}.*

As a consequence of the Theorem 5.1, we can show the ergodicity of B-connected random chains in expectation.

Corollary 5.1 *Let $\{W(k)\}$ be an independent random chain with weak feedback property. Also, suppose that $\{\bar{W}(k)\}$ is B-connected. Then, $\{W(k)\}$ is ergodic almost surely.*

Proof By Theorem 2.4, for any such a chain, $\{\bar{W}(k)\}$ is an ergodic chain. Furthermore, by Theorem 2.4, we have $\lim_{k\to\infty} \bar{W}(k : t_0) = ev^T(t_0)$ where $v(t_0) \geq \eta > 0$ for some $\eta > 0$. Thus, $\{v(k)\} \geq \eta$ is a uniformly bounded absolute probability sequence for $\{\bar{W}(k)\}$. Thus, by Theorem 5.1, $\{W(k)\}$ is ergodic almost surely. **Q.E.D.**

5.1.1 Rate of Convergence

Here we establish a rate of convergence result for independent random chains in $\mathcal{P}^*$ with infinite flow property and with weak feedback property.

To derive the rate of convergence result, we first show some intermediate results.

Lemma 5.3 *Let $\{A(k)\}$ be a chain of stochastic matrices and $\{z(k)\}$ be the dynamics driven by $\{A(k)\}$ started at time $t_0 = 0$ with a starting point $z(0) \in \mathbb{R}^m$. Let σ be a permutation of the index set $[m]$ corresponding to the nondecreasing ordering of the entries $z_\ell(0)$, i.e., σ is a permutation on $[m]$ such that $z_{\sigma_1}(0) \leq \cdots \leq z_{\sigma_m}(0)$. Also, let $T \geq 1$ be such that*

$$\sum_{k=0}^{T-1} A_S(k) \geq \delta \quad for\ every\ non\text{-}trivial\ S \subset [m], \tag{5.1}$$

where $\delta \in (0, 1)$ is arbitrary. Then, we have

$$\sum_{k=0}^{T-1} \sum_{i<j} \left(A_{ij}(k) + A_{ji}(k)\right) (z_j(k) - z_i(k))^2 \geq \frac{\delta(1-\delta)^2}{z_{\sigma_m}(0) - z_{\sigma_1}(0)} \sum_{i=1}^{m-1} (z_{\sigma_{i+1}}(0) - z_{\sigma_i}(0))^3.$$

Proof Relation (5.1) holds for any nontrivial set $S \subset [m]$. Hence, without loss of generality we may assume that the permutation σ is identity (otherwise we will relabel the indices of the entries in $z(0)$ and update the matrices accordingly). Thus, we have $z_1(0) \leq \cdots \leq z_m(0)$. For each $\ell = 1, \ldots, m-1$, let $S_\ell = \{1, \ldots, \ell\}$ and define time $t_\ell \geq 1$, as follows:

$$t_\ell = \operatorname*{argmin}_{t \geq 1} \left\{ \sum_{k=0}^{t-1} A_{S_\ell}(k) \geq \delta \frac{z_{\ell+1}(0) - z_\ell(0)}{z_m(0) - z_1(0)} \right\}.$$

Since the entries of $z(0)$ are nondecreasing, we have $\delta \frac{z_{\ell+1}(0)-z_\ell(0)}{(z_m(0)-z_1(0))} \leq \delta$ for all $\ell = 1, \ldots, m-1$. Thus, by relation (5.1), the time $t_\ell \geq 1$ exists and $t_\ell \leq T$ for each ℓ.

We next estimate $z_j(k) - z_i(k)$ for all $i < j$ and any time $k = 0, \cdots, T-1$. For this, we introduce for $0 \leq k \leq T-1$ and $i < j$ the index sets $a_{ij}(k) \subset [m]$, as follows:

$$a_{ij}(k) = \{\ell \in [m] \mid k \leq t_\ell - 1, \ \ell \geq i, \ \ell + 1 \leq j\}.$$

Let $k \leq t_\ell - 1$ for some ℓ. Since $S_\ell = \{1, \ldots, \ell\}$, we have $i \in S_\ell$ and $j \in \bar{S}_\ell$. Thus, by Lemma 4.9, for any $k \geq 1$, we have

$$z_i(k) \leq \max_{s \in S_\ell} z_s(0) + (z_m(0) - z_1(0)) \sum_{\tau=0}^{k-1} A_{S_\ell}(\tau),$$

$$z_j(k) \geq \min_{r \in \bar{S}_\ell} z_r(0) - (z_m(0) - z_1(0)) \sum_{\tau=0}^{k-1} A_{S_\ell}(\tau).$$

Furthermore, $\max_{s \in S_\ell} z_s(0) = z_\ell(0)$ and $\min_{r \in \bar{S}_\ell} z_r(0) = z_{l+1}(0)$ since $S_\ell = \{1, \ldots, \ell\}$ and $z_1(0) \leq \cdots \leq z_m(0)$. Thus, it follows

$$z_i(k) - z_\ell(0) \leq (z_m(0) - z_1(0)) \sum_{\tau=0}^{k-1} A_{S_\ell}(\tau),$$

$$z_{\ell+1}(0) - z_j(k) \leq (z_m(0) - z_1(0)) \sum_{\tau=0}^{k-1} A_{S_\ell}(\tau).$$

By the definition of time t_ℓ, we have $(z_m(0) - z_1(0)) \sum_{\tau=0}^{k-1} A_{S_\ell}(\tau) < \delta(z_{\ell+1}(0) - z_\ell(0))$ for $k \le t_\ell - 1$. Hence, by using this and the definition of $a_{ij}(k)$, for any $\ell \in a_{ij}(k)$ we have

$$z_i(k) - z_\ell(0) \le \delta(z_{\ell+1}(0) - z_\ell(0)), \tag{5.2}$$

$$z_{\ell+1}(0) - z_j(k) \le \delta(z_{\ell+1}(0) - z_\ell(0)). \tag{5.3}$$

Now suppose that $a_{ij}(k) = \{\ell_1, \ldots, \ell_r\}$ for some $r \le m - 1$ and $\ell_1 \le \cdots \le \ell_r$. By choosing $\ell = \ell_1$ in (5.2) and $\ell = \ell_r$ in (5.3), and by letting $\alpha_i = z_{i+1}(0) - z_i(0)$, we obtain

$$z_j(k) - z_i(k) \ge z_{\ell_r+1}(0) - z_{\ell_1}(0) - \delta(\alpha_{\ell_r} + \alpha_{\ell_1}).$$

Since $z_i(0) \le z_{i+1}(0)$ for all $i = 1, \ldots, m - 1$, we have $z_{\ell_1}(0) \le z_{\ell_1+1}(0) \le \cdots \le z_{\ell_r}(0) \le z_{\ell_r+1}(0)$, which combined with the preceding relation yields $z_j(k) - z_i(k) \ge \sum_{\xi=1}^{r}(z_{\ell_\xi+1}(0) - z_{\ell_\xi}(0)) - \delta(\alpha_{\ell_r} + \alpha_{\ell_1})$. Using $\alpha_i = z_{i+1}(0) - z_i(0)$ and $a_{ij}(k) = \{\ell_1, \ldots, \ell_r\}$, we further have

$$z_j(k) - z_i(k) \ge \sum_{\xi=1}^{r} \alpha_{\ell_\xi} - \delta(\alpha_{\ell_r} + \alpha_{\ell_1}) \ge (1-\delta) \sum_{\xi=1}^{r-1} \alpha_{\ell_\xi} = (1-\delta) \sum_{\ell \in a_{ij}(k)} \alpha_\ell. \tag{5.4}$$

By Eq. (5.4), it follows that

$$\sum_{i<j} \left(A_{ij}(k) + A_{ji}(k)\right)(z_j(k) - z_i(k))^2 \ge (1 - \delta)^2 \sum_{i<j} \left(A_{ij}(k) + A_{ji}(k)\right) \left(\sum_{\ell \in a_{ij}(k)} \alpha_\ell\right)^2$$

$$\ge (1 - \delta)^2 \sum_{i<j} \left(A_{ij}(k) + A_{ji}(k)\right) \left(\sum_{\ell \in a_{ij}(k)} \alpha_\ell^2\right),$$

where the last inequality holds by $\alpha_\ell \ge 0$. In the last term in the preceding relation, the coefficient of α_ℓ^2 is equal to $(1 - \delta)^2 A_{S_\ell}(k)$. Furthermore, by the definition of $a_{ij}(k)$, we have $\ell \in a_{ij}(k)$ only when $k \le t_\ell - 1$. Therefore,

$$\sum_{i<j} \left(A_{ij}(k) + A_{ji}(k)\right)(z_j(k) - z_i(k))^2$$

$$\ge (1 - \delta)^2 \sum_{i<j} \left(A_{ij}(k) + A_{ji}(k)\right) \left(\sum_{\ell \in a_{ij}(k)} \alpha_\ell^2\right) = (1 - \delta)^2 \sum_{\{\ell \mid k \le t_\ell - 1\}} A_{S_\ell}(k)\alpha_\ell^2.$$

Summing these relations over $k = 0, \ldots, T - 1$, we obtain

$$\sum_{k=0}^{T-1} \sum_{i<j} \left(A_{ij}(k) + A_{ji}(k) \right) (z_j(k) - z_i(k))^2 \geq (1-\delta)^2 \sum_{k=0}^{T-1} \sum_{\{\ell \mid k \leq t_\ell - 1\}} A_{S_\ell}(k) \alpha_\ell^2$$

$$\geq (1-\delta)^2 \sum_{\ell=1}^{m-1} \left(\sum_{k=0}^{t_\ell - 1} A_{S_\ell}(k) \right) \alpha_\ell^2,$$

where the last inequality follows by exchanging the order of summation. By the definition of t_ℓ and using $\alpha_\ell = z_{\ell+1}(0) - z_\ell(0)$, we have $\sum_{k=0}^{t_\ell - 1} A_{S_\ell}(k) \geq \frac{\delta \alpha_\ell}{z_m(0) - z_1(0)}$, implying

$$\sum_{k=0}^{T-1} \sum_{i<j} \left(A_{ij}(k) + A_{ji}(k) \right) (z_j(k) - z_i(k))^2 \geq \delta(1-\delta)^2 \sum_{\ell=1}^{m-1} \frac{\alpha_\ell^3}{z_m(0) - z_1(0)}.$$

Q.E.D.

Another intermediate result that we use in our forthcoming discussions is the following.

Lemma 5.4 *Let $\pi \in \mathbb{R}^m$ be a stochastic vector, and let $x \in \mathbb{R}^m$ be such that $x_1 \leq \cdots \leq x_m$. Then, we have*

$$\frac{1}{m-1} V(x) \leq \sum_{i=1}^{m-1} (x_{i+1} - x_i)^2, \tag{5.5}$$

$$\sum_{i=1}^{m-1} \frac{(x_{i+1} - x_i)^2}{m-1} \leq \sum_{i=1}^{m-1} \frac{(x_{i+1} - x_i)^3}{x_m - x_1}. \tag{5.6}$$

and hence,

$$\frac{1}{(m-1)^2} V(x) \leq \frac{1}{x_m - x_1} \sum_{i=1}^{m-1} (x_{i+1} - x_i)^3,$$

where $V(x) = \sum_{i=1}^m \pi_i (x_i - \pi^T x)^2$.

Proof We first show relation (5.5). We have $x_i \leq x_m$ for all i. Since π is stochastic, we also have $x_m \geq \pi^T x \geq x_1$. Thus, $V(x) = \sum_{i=1}^m \pi_i (x_i - \pi^T x)^2 \leq (x_m - x_1)^2$. By writing $x_m - x_1 = \sum_{i=1}^{m-1} (x_{i+1} - x_i)$, we obtain

$$(x_m - x_1)^2 = (m-1)^2 \left(\frac{1}{m-1} \sum_{i=1}^{m-1} (x_{i+1} - x_i) \right)^2 \leq (m-1) \sum_{i=1}^{m-1} (x_{i+1} - x_i)^2,$$

where the last inequality holds by the convexity of the function $s \mapsto s^2$. Using $V(x) \leq (x_m - x_1)^2$ and the preceding relation we obtain relation (5.5).

To prove relation (5.6), we have

$$(x_m - x_1) \sum_{i=1}^{m-1} (x_{i+1} - x_i)^2 = \sum_{j=1}^{m-1} (x_{j+1} - x_j) \sum_{i=1}^{m-1} (x_{i+1} - x_i)^2$$

$$= \sum_{j=1}^{m-1} (x_{j+1} - x_j)^3 + \sum_{j<i} \left((x_{j+1} - x_j)(x_{i+1} - x_i)^2 + (x_{i+1} - x_i)(x_{j+1} - x_j)^2 \right)$$

$$\tag{5.7}$$

Now, consider scalars $\alpha \geq 0$ and $\beta \geq 0$, and let $u = (\alpha, \beta)$ and $v = (\beta^2, \alpha^2)$. Then, by Hölder's inequality (Theorem B.1) with $p = 3$, $q = \frac{3}{2}$, we have $u^T v \leq \|u\|_p \|v\|_q$. Hence,

$$\alpha\beta^2 + \beta\alpha^2 \leq \left(\alpha^3 + \beta^3 \right)^{\frac{1}{3}} \left(\beta^3 + \alpha^3 \right)^{\frac{2}{3}} = \alpha^3 + \beta^3. \tag{5.8}$$

By using (5.8) in (5.7) with $\alpha_j = (x_{j+1} - x_j)$ and $\beta_i = (x_{i+1} - x_i)$ for different indices j and i, $1 \leq j < i \leq m - 1$, we obtain

$$(x_m - x_1) \sum_{i=1}^{m-1} (x_{i+1} - x_i)^2 \leq \sum_{j=1}^{m-1} (x_{j+1} - x_j)^3 + \sum_{j<i} \left((x_{j+1} - x_j)^3 + (x_{i+1} - x_i)^3 \right)$$

$$= (m - 1) \sum_{i=1}^{m-1} (x_{i+1} - x_i)^3,$$

which completes the proof. **Q.E.D.**

Now, let $\{W(k)\}$ be an independent random chain with infinite flow property. Let $t_0 = 0$ and for any $q \geq 1$, let

$$t_q = \underset{t \geq t_{q-1}+1}{\text{argmin}} \ \Pr\left(\min_{S \subset [m]} \sum_{k=t_{q-1}}^{t-1} W_S(k) \geq \delta \right) \geq \epsilon, \tag{5.9}$$

where $\epsilon, \delta \in (0, 1)$ are arbitrary. Define

$$\mathscr{A}_q = \left\{ \omega \ \middle| \ \min_{S \subset [m]} \sum_{k=t_q}^{t_{q+1}-1} W_S(k, \omega) \geq \delta \right\} \quad \text{for } q \geq 0. \tag{5.10}$$

Since the chain has infinite flow property, the infinite flow event $\mathscr{F}$ occurs a.s. Therefore, the time t_q is finite for all q.

We next provide a rate of convergence result for a random chain in $\mathcal{P}^*$ with weak feedback property.

Theorem 5.2 *Let $\{W(k)\}$ be an independent random chain in $\mathcal{P}^*$ and weak feedback property. Then, for any $q \geq 1$, we have:*

$$E\big[V(x(t_q), t_q)\big] \leq \left(1 - \frac{\epsilon \delta(1-\delta)^2 \gamma p^*}{(m-1)^2}\right)^q E\big[V(x(0), 0)\big]$$

Proof Let $\{\pi(k)\} \geq p^*$ be an absolute probability sequence for $\{W(k)\}$. Fix $v \in \mathbb{R}^m$ and let $x(0, \omega) = v$ for any $\omega \in \Omega$. Let us denote the (random) ordering of the entries of the random vector $x(t_q)$ by η^q for all q. Thus, at time t_q, we have $x_{\eta_1^q}(t_q) \leq \cdots \leq x_{\eta_m^q}(t_q)$.

Now, let $q \geq 0$ be arbitrary and fixed, and consider the set $\mathscr{A}_q$ in (5.10). By the definition of $\mathscr{A}_q$, we have $\sum_{k=t_q}^{t_{q+1}-1} W_S(k, \omega) \geq \delta$ for any $S \subset [m]$ and $\omega \in \mathscr{A}_q$. Thus, by Lemma 5.3, we obtain for any $\omega \in \mathscr{A}_q$,

$$\sum_{k=t_q}^{t_{q+1}-1} \sum_{i<j} \left(W_{ij}(k) + W_{ji}(k)\right)(x_i(k) - x_j(k))^2(\omega) \geq \frac{\delta(1-\delta)^2}{d(t_q)(\omega)} \sum_{\ell=1}^{m-1}(x_{\eta_{\ell+1}^q}(t_q) - x_{\eta_\ell^q}(t_q))^3(\omega)$$

$$\geq \frac{\delta(1-\delta)^2}{(m-1)^2} V(x(t_q), t_q)(\omega),$$

where $d(t_q) = x_{\eta_m^q}(t_q) - x_{\eta_1^q}(t_q)$ and the last inequality follows by Lemma 5.4. We can compactly write the inequality as:

$$\sum_{k=t_q}^{t_{q+1}-1} \sum_{i<j} \left(W_{ij}(k) + W_{ji}(k)\right)(x_i(k) - x_j(k))^2 \geq \frac{\delta(1-\delta)^2}{(m-1)^2} V(x(t_q), t_q)\mathbf{1}_{\mathscr{A}_q},$$

$$(5.11)$$

Observe that $x(k)$ and $W(k)$ are independent since the chain is independent. Therefore, by Theorem 4.4, we have

$$E\big[V(x(t_{q+1}), t_{q+1}) - V(x(t_q), t_q)\big] \leq - \sum_{k=t_q}^{t_{q+1}-1} \sum_{i<j} H_{ij}(k)E\Big[(x_i(k) - x_j(k))^2\Big],$$

with $H(k) = E\big[W^T(k)\mathrm{diag}(\pi(k+1))W(k)\big]$. Since $\{\pi(k)\} \geq \mathcal{P}^* > 0$, we have,

$$H_{ij}(k) = E\Big[(W^i(k))^T \mathrm{diag}(\pi(k+1))W^j(k)\Big] \geq p^* E\Big[(W^i(k))^T W^j(k)\Big]$$

$$\geq p^* \gamma \left(E\big[W_{ij}(k)\big] + E\big[W_{ji}(k)\big]\right).$$

Therefore,

$$E\big[V(x(t_{q+1}))\big] - E\big[V(x(t_q))\big] \leq -p^* \gamma E\left[\sum_{k=t_q}^{t_q-1} \sum_{i<j} \left(W_{ij}(k) + W_{ji}(k)\right)(x_i(k) - x_j(k))^2\right].$$

Further, using relation (5.11), we obtain

$$\mathsf{E}\big[V(x(t_{q+1}),t_{q+1})\big] - \mathsf{E}\big[V(x(t_q),t_q)\big] \leq -p^*\gamma\,\mathsf{E}\left[\frac{\delta(1-\delta)^2}{(m-1)^2}\,\mathbf{1}_{A_q}V(x(t_q),t_q)\right]$$

$$\leq -\frac{\epsilon\delta(1-\delta)^2\gamma p^*}{(m-1)^2}\,\mathsf{E}\big[V(x(t_q),t_q)\big],$$

where the last inequality follows by $\Pr\left(\mathscr{A}_q\right) \geq \epsilon$, and the fact that $\mathbf{1}_{\mathscr{A}_q}$ and $V(x(t_q),t_q)$ are independent (since $x(t_q)$ depends on information prior to time t_q and the set $\mathscr{A}_q$ relies on information at time t_q and later). Hence, it follows

$$\mathsf{E}\big[V(x(t_{q+1}),t_{q+1})\big] \leq \left(1 - \frac{\epsilon\delta(1-\delta)^2\gamma p^*}{(m-1)^2}\right)\mathsf{E}\big[V(x(t_q),t_q)\big].$$

Therefore, for arbitrary $q \geq 0$ we have

$$\mathsf{E}\big[V(x(t_q),t_q)\big] \leq \left(1 - \frac{\epsilon\delta(1-\delta)^2\gamma p^*}{(m-1)^2}\right)^q \mathsf{E}\big[V(x(0),0)\big].$$

Q.E.D.

Note that in Theorem 5.2, the starting point $x(0)$ can be random. However, if $x(0) = v$ a.s. for some $v \in \mathbb{R}^m$, then $\mathsf{E}[V(x(0),0)] = V(v,0)$. Also, if we let $d(x) = \max_{i\in[m]} x_i - \min_{j\in[m]} x_j$, then $|v_i - \pi^T(0)v| \leq d(v)$ for all $i \in [m]$, and hence,

$$\mathsf{E}[V(x(0),0)] = V(v,0) = \sum_{i=1}^{m} \pi_i(0)(v_i - \pi^T(0)v)^2 \leq \sum_{i=1}^{m} \pi_i(0)d^2(v) = d^2(v),$$

where the last equality follow from $\pi(0)$ being stochastic. Thus, the following corollary is immediate.

Corollary 5.2 *Let $\{W(k)\}$ be an independent random chain in $\mathcal{P}^*$ with weak feedback property, started at a deterministic point $v \in \mathbb{R}^m$. Then, for any $q \geq 1$, we have*:

$$\mathsf{E}\big[V(x(t_q),t_q)\big] \leq \left(1 - \frac{\epsilon\delta(1-\delta)^2\gamma p^*}{(m-1)^2}\right)^q d^2(v).$$

5.1.2 i.i.d. Chains

As we assume more structure on a random chain, we may deduce more of its properties. For example, as discussed earlier in this chapter, for any independent random chain, the ergodicity and the infinite flow property are trivial events. However, the consensus event may not be a trivial event as shown in the following example.

Example 5.1 For $p \in (0, 1)$, let $\{W(k)\}$ be an independent random chain defined by

$$W(0) = \begin{cases} \frac{1}{m}J & \text{with probability } p \\ I & \text{with probability } 1 - p \end{cases},$$

and $W(k) = I$ for all $k > 0$. Then, although the chain $\{W(k)\}$ is an independent chain, it admits consensus with probability p and hence, the consensus event is not a trivial event for this chain.

Although the consensus event is not a trivial event for an independent chain, it is a trivial event for an i.i.d. chain. To establish this, we make use of the following lemma.

Lemma 5.5 *Let $A \in \mathbb{S}_m$ and $x \in \mathbb{R}^m$. Also, let A be such that*

$$\max_{i \in [m]}[Ae_\ell]_i - \min_{j \in [m]}[Ae_\ell]_j \leq \frac{1}{2m} \text{ for any } \ell \in [m],$$

where $[v]_i$ denotes the ith component of a vector v. Then, we have $\max_i[Ax]_i - \min_j[Ax]_j \leq \frac{1}{2}$ for any $x \in [0, 1]^m$.

Proof Let $x \in \mathbb{R}^m$ with $x_\ell \in [0, 1]$ for any $\ell \in [m]$. Then, we have for any $i, j \in [m]$,

$$y_i - y_j = \sum_{\ell=1}^{m}(A_{i\ell} - A_{j\ell})x_\ell \leq \sum_{\ell=1}^{m}|A_{i\ell} - A_{j\ell}| = \sum_{\ell=1}^{m}\left|[Ae_\ell]_i - [Ae_\ell]_j\right|.$$

By the assumption on A, we obtain $\left|[Ae_\ell]_i - [Ae_\ell]_j\right| \leq \max_{i \in [m]}[Ae_\ell]_i - \min_{j \in [m]}[Ae_\ell]_j \leq \frac{1}{2m}$. Hence, $y_i - y_j \leq \sum_{\ell=1}^{m}\frac{1}{2m} = \frac{1}{2}$, implying $\max_i y_i - \min_j y_j \leq \frac{1}{2}$. **Q.E.D.**

We now provide our main result for i.i.d. chains, which states that the ergodicity and the consensus events are almost surely equal for such chains. We establish this result by using Lemma 5.5 and the Borel-Cantelli lemma (Lemma A.6).

Theorem 5.3 *We have $\mathscr{E} = \mathscr{C}$ almost surely for any i.i.d. random chain.*

Proof Since $\mathscr{E} \subseteq \mathscr{C}$, the assertion is true when consensus occurs with probability 0. Therefore, it suffices to show that if the consensus occurs with a probability p other than 0, the two events are almost surely equal. Let $\Pr(\mathscr{C}) = p$ with $p \in (0, 1]$. Then, for all $\omega \in \mathscr{C}$,

$$\lim_{k \to \infty} d(x(k, \omega)) = 0,$$

where $d(x) = \max_i x_i - \min_j x_j$ and $\{x(k, \omega)\}$ is the sequence generated by the dynamic system (3.1) with some $x(0, \omega) \in \mathbb{R}^m$.

For every $\ell \in [m]$, let $\{x^\ell(k, \omega)\}$ be the sequence generated by the dynamic system in (3.1) with $x(0, \omega) = e_\ell$. Then, for any $\omega \in \mathscr{C}$, there is the smallest integer $K^\ell(\omega) \geq 0$ such that

$$d(x^\ell(k, \omega)) \leq \frac{1}{2m} \qquad \text{for all } k \geq K^\ell(\omega).$$

Note that $d(x^\ell(k, \omega))$ is a non-increasing sequence (of k) for each $\ell \in [m]$. Hence, by letting $K(\omega) = \max_{\ell \in [m]} K^\ell(\omega)$ we obtain $d(x^\ell(k, \omega)) \leq \frac{1}{2m}$ for all $\ell \in [m]$ and $k \geq K(\omega)$. Thus, by applying Lemma 5.5, we have for almost all $\omega \in \mathscr{C}$,

$$d(x(k, \omega)) \leq \frac{1}{2}, \tag{5.12}$$

for all $k \geq K(\omega)$ and $x(0) \in [0, 1]^m$. By the definition of consensus, we have

$$\lim_{N \to \infty} \Pr(K \leq N) \geq \Pr(\mathscr{C}) = p.$$

Thus, by the continuity of the measure, there exists an integer N_1 such that $\Pr(K < N_1) \geq \frac{p}{2}$.

Now, let time $T \geq 0$ be arbitrary, and let l_k^T denote the N_1-tuple of the random matrices $W(s)$ driving the system (3.1) for $s = T + N_1 k, \ldots, T + N_1(k + 1) - 1$ and $k \geq 0$, i.e.,

$$l_k^T = \Big(W(T + N_1 k), W(T + N_1 k + 1), \ldots, W(T + N_1(k + 1) - 1) \Big) \quad \text{for all } k \geq 0.$$

Let L_N denote the collection of all N-tuples $(A_1, \ldots, A_N)$ of matrices $A_i \in \mathbb{S}_m$, $i \in [N]$ such that for $x(N) = A_N A_{N-1} \cdots A_1 x(0)$ with $x(0) \in [0, 1]^m$, we have $d(x(N)) \leq \frac{1}{2}$. By the definitions of l_k^T and L_N, relation (5.12) and relation $\Pr(K < N_1) \geq \frac{p}{2}$ state that $\Pr\left(\{l_0^T \in L_{N_1}\}\right) \geq \frac{p}{2}$. By the i.i.d. property of the chain, the events $\{l_k^T \in L_{N_1}\}$, $k \geq 0$, are i.i.d. and the probability of their occurrence is equal to $\Pr\left(\{l_0^T \in L_{N_1}\}\right)$, implying that $\Pr\left(\{l_k^T \in L_{N_1}\}\right) \geq \frac{p}{2}$ for all $k \geq 0$. Consequently, $\sum_{k=0}^{\infty}[1]\Pr\left(\{l_k^T \in L_{N_1}\}\right) = \infty$. Since the events $\{l_k^T \in L_{N_1}\}$ are i.i.d., by Borel-Cantelli lemma (Lemma A.6) $\Pr\left(\{\omega \in \Omega \mid \omega \in \{l_k^T \in L_{N_1}\} i.o.\}\right) = 1$. Observing that the event $\{\omega \in \Omega \mid \omega \in \{l_k^T \in L_{N_1}\} \ i.o.\}$ is contained in the consensus event for the chain $\{W(T + k)\}_{k \geq 0}$, we see that the consensus event for the chain $\{W(T + k)\}_{k \geq 0}$ occurs almost surely. Since this is true for arbitrary $T \geq 0$, it follows that the chain $\{W(k)\}$ is ergodic $a.s.$, implying $\mathscr{C} \subseteq \mathscr{E}$ $a.s.$ This and the inclusion $\mathscr{E} \subseteq \mathscr{C}$ yield $\mathscr{C} = \mathscr{E}$ $a.s.$ **Q.E.D.**

Theorem 5.3 extends the equivalence result between the consensus and ergodicity for i.i.d. chains given in Theorem 3.a and Theorem 3.b of [1] (and hence Corollary 4 in [1]), which have been established there assuming that the matrices have positive diagonal entries almost surely.

The relations among $\mathscr{C}$, $\mathscr{E}$, and $\mathscr{F}$ for i.i.d. random chains are illustrated in Fig. 5.1.

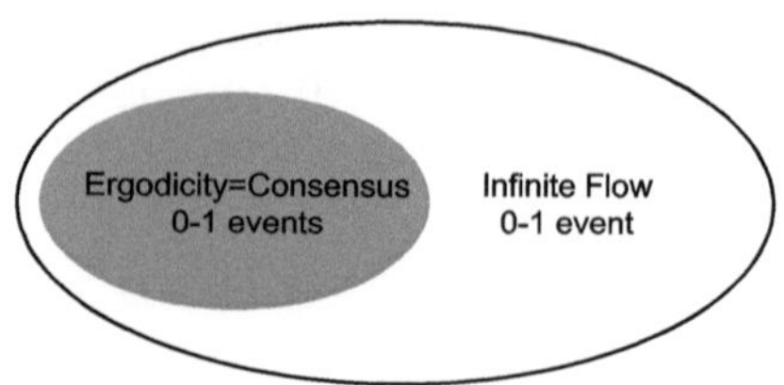

Fig. 5.1 Relations among consensus, ergodicity, and infinite flow events for an i.i.d. chain

5.2 Non-Negative Matrix Theory

In this section, we show that Theorem 4.10 is a generalization of a well-known result in the non-negative matrix theory which plays a central role in the theory of ergodic Markov chains. For this let us revisit the definitions of aperiodicity and irreducibility for a stochastic matrix.

Definition 5.1 (*Irreducibility*, [2, p. 45]) A stochastic matrix A is said to be irreducible if there is no permutation matrix P such that:

$$P^T A P = \begin{bmatrix} X & Y \\ \mathbf{0} & Z \end{bmatrix},$$

where X, Y, Z are $i \times i$, $i \times (m - i)$, and $(m - i) \times (m - i)$ matrices for some $i \in [m - 1]$ and $\mathbf{0}$ is the $(m - i) \times i$ matrix with all entries equal to zero.

Definition 5.2 (*Aperiodicity*, [3, p. 119]) An irreducible matrix A is said to be aperiodic if we have

$$g.c.d. \left(\{n \mid A_{ii}^n > 0\} \right) = 1 \quad for\ some\ i \in [m],$$

where $g.c.d.(M)$ is the greatest common divisor of the elements in the set $M \subseteq \mathbb{Z}^+$. Furthermore, it is said to be strongly aperiodic if $A_{ii} > 0$ for all $i \in [m]$.

For an aperiodic and irreducible matrix A, we have the following result.

Lemma 5.6 [2, p. 46] *Let A be an irreducible and aperiodic stochastic matrix. Then, A^k converges to a rank matrix.*

Let us reformulate irreducibility using the tools we have developed in Chaps. 3 and 4.

Lemma 5.7 *A stochastic matrix A is an irreducible matrix if and only if the static chain $\{A\}$ is a balanced chain with infinite flow property.*

Proof By the definition, a matrix A is irreducible if there is no permutation matrix P such that

$$P^T A P = \begin{bmatrix} X & Y \\ \mathbf{0} & Z \end{bmatrix}.$$

Since $A \geq 0$, we have that A is reducible if and only if there exists a subset $S = \{1, \ldots, i\}$ for some $i \in [m-1]$, such that

$$0 = [P^T A P]_{\bar{S}S} = \sum_{i \in \bar{S}, j \in S} e_i [P^T A P] e_j = \sum_{i \in \bar{S}, j \in S} A_{\sigma_i \sigma_j} = \sum_{i \in \bar{Q}, j \in Q} A_{ij},$$

where $\sigma_i = P(\{i\})$ and $Q = \{\sigma_i \mid i \in S\}$. Thus, A is irreducible if and only if $A_{S\bar{S}} > 0$ for any non-trivial $S \subset [m]$. Therefore, if we let

$$\alpha = \min_{\substack{S \subset [m] \\ S \neq \emptyset}} \frac{A_{S\bar{S}}}{A_{\bar{S}S}},$$

then $\alpha > 0$ and we conclude that $\{A\}$ is balanced with balanced-ness coefficient α. Also since $A_S \geq A_{S\bar{S}} > 0$, we conclude that $\{A\}$ has infinite flow property.

Now, if $\{A\}$ has infinite flow property, then it follows that $A_{S\bar{S}} > 0$ or $A_{\bar{S}S} > 0$ for any non-trivial $S \subset [m]$. By balanced-ness, it follows that $\min(A_{S\bar{S}}, A_{\bar{S}S}) > 0$ for any non-trivial $S \subset [m]$, implying that A is irreducible. **Q.E.D.**

For the aperiodicity, note that A is strongly aperiodic if and only if $\{A\}$ has strong feedback property. Thus, let us generalize the concepts of irreducibility and strong aperiodicity to an *independent random chain*.

Definition 5.3 We say that an independent random chain $\{W(k)\}$ is irreducible if $\{\bar{W}(k)\}$ is a balanced chain with infinite flow property. Furthermore, we say that $\{W(k)\}$ is strongly aperiodic if $\{W(k)\}$ has feedback property.

Based on Definition 5.3 and Theorem 4.5, we have the following extension of Lemma 5.6 for random chains.

Theorem 5.4 *Let $\{W(k)\}$ be an irreducible and strongly aperiodic independent random chain. Then, for any $t_0 \geq 0$, the product $W(k : t_0)$ converges to a rank one stochastic matrix almost surely (as k goes to infinity). Moreover, if $\{W(k)\}$ does not have the infinite flow property, the product $W(k : t_0)$ almost surely converges to a (random) matrix that has rank at most τ for any $t_0 \geq 0$, where τ is the number of connected components of the infinite flow graph of $\{\bar{W}(k)\}$.*

Proof By Theorem 4.5, any balanced chain $\{\bar{W}(k)\}$ with strong feedback property is infinite flow stable. Also, by Lemma 5.2 $\{\bar{W}(k)\}$ has infinite flow property if and only if $\{W(k)\}$ has infinite flow property. Thus we conclude that $\{W(k)\}$ is ergodic almost surely which means that for any $t_0 \geq 0$, the product $W(k : t_0)$ converges almost surely to a rank one stochastic matrix as k goes to infinity.

Now, if $\{W(k)\}$ does not have the infinite flow property, by Theorem 5.1, for any $t_0 \geq 0$, the product $W(k : t_0)$ converges to some matrix $W(\infty : t_0)$. Also, $\lim_{k \to \infty} \|W_i(k : t_0) - W_j(k : t_0)\| = 0$ for any i,j belonging to a same connected component of the infinite flow graph of $\{\bar{W}(k)\}$. Thus, the rows of $W(\infty : t_0)$ admits at most τ different values, where τ is the number of the connected components of the infinite flow graph of $\{\bar{W}(k)\}$. **Q.E.D.**

An immediate consequence of Theorem 5.4 is a generalization of Lemma 5.6 to inhomogeneous chains.

Corollary 5.3 *Let* $\{A(k)\}$ *be an irreducible chain of stochastic matrices that is strongly aperiodic. Then,* $A(\infty : t_0) = \lim_{k \to \infty} A(k : t_0)$ *exists and it is a rank one matrix for any* $t_0 \geq 0$.

5.3 Convergence Rate for Uniformly Bounded Chains

Consider a deterministic chain $\{A(k)\}$ that is uniformly bounded, i.e. $A_{ij}(k) \geq \gamma$ for any $i, j \in [m]$ and $k \geq 0$ such that $A_{ij}(k) > 0$, where $\gamma > 0$. Here, we provide a rate of convergence result for a uniformly bounded chain that, as special cases, includes several of the existing results that have been derived using different methods.

Let $\{A(k)\}$ be a deterministic chain that is uniformly bounded by $\gamma > 0$. By the necessity of the infinite flow (Theorem 3.1), to ensure ergodicity, $\{A(k)\}$ should have infinite flow property. So as in the case of random dynamics [Eq. (5.9)], let $t_0 = 0$ and for $q \geq 1$, recursively, let t_q be defined by

$$t_q = \underset{t \geq t_{q-1}+1}{\operatorname{argmin}} \ \underset{S \subset [m]}{\min} \ \sum_{k=t_{q-1}}^{t-1} A_S(k) > 0. \tag{5.13}$$

Basically, t_q is the qth time that there is a non-zero flow over every cut $[S, \bar{S}]$, for every non-trivial $S \subset [m]$. Here, using the same line of argument as in the proof of Lemma 9 in [4], we have the following result.

Theorem 5.5 *Let* $\{A(k)\}$ *be a chain in* $\mathcal{P}^*$ *with feedback property and let* t_q *be as defined in Eq. (5.13). Then, for any dynamics* $\{x(k)\}$ *driven by* $\{A(k)\}$ *and for any* $q \geq 0$, *we have*

$$V_\pi(x(t_{q+1}), t_{q+1}) \leq \left(1 - \frac{\gamma p^*}{2(m-1)} \right) V_\pi(x(t_q), t_q),$$

where $\mathcal{P}^* > 0$ *is such that* $\{\pi(k)\} \geq \mathcal{P}^*$ *for an absolute probability sequence* $\{\pi(k)\}$ *of* $\{A(k)\}$.

Proof Note that, since $\{A(k)\}$ has feedback property and is uniformly bounded, it follows that $A_{ii}(k) \geq \gamma$ for any $k \geq 0$ and $i \in [m]$. Now, by Theorem 4.4, for any $q \geq 0$, we have:

$$V_\pi(x(t_{q+1}), t_{q+1}) = V_\pi(x(t_q), t_q) - \sum_{k=t_q}^{t_{q+1}-1} \sum_{i<j} H_{ij}(k)(x_i(k) - x_j(k))^2$$

$$\leq V_\pi(x(t_q), t_q) - \mathcal{P}^* \sum_{k=t_q}^{t_{q+1}-1} \sum_{i<j} L_{ij}(k)(x_i(k) - x_j(k))^2, \tag{5.14}$$

where $H(k) = A^T(k)\mathrm{diag}(\pi(k+1))A(k)$ and $L(k) = A^T(k)A(k)$. Now, without loss of generality we can assume that $x(t_q)$ is ordered, i.e. $x_1(t_q) \leq \cdots \leq x_m(t_q)$ and following the same lines of arguments as those used to derive Lemma 8 in [4], it follows that

$$V_\pi(x(t_q), t_q) - V(x(t_{q+1}), t_{q+1}) \geq \frac{\gamma \mathcal{P}^*}{2} \sum_{i=1}^{m-1} (x_{i+1}(t_q) - x_i(t_q))^2. \tag{5.15}$$

But by Lemma 5.4, it follows that

$$\sum_{i=1}^{m-1} (x_{i+1}(t_q) - x_i(t_q))^2 \geq \frac{1}{m-1} V_\pi(x(t_q), t_q). \tag{5.16}$$

Combining (5.15) and (5.16), we conclude

$$V_\pi(x(t_q), t_q) - V_\pi(x(t_{q+1}), t_{q+1}) \geq \frac{\gamma p^*}{2(m-1)} V_\pi(x(t_q), t_q),$$

which completes the proof. **Q.E.D.**

Now, if we have B-connectivity assumption on the chain $\{A(k)\}$, we can replace t_q by qB and hence, the following result follows immediately.

Corollary 5.4 *For a B-connected chain $\{A(k)\}$ with an absolute probability sequence $\{\pi(k)\} \geq p^*$, we have*: **Q.E.D.**

$$V_\pi(x((q+1)B), (q+1)B) \leq \left(1 - \frac{\gamma p^*}{2(m-1)}\right) V_\pi(x(qB), qB) \qquad \text{for all } q \geq 0,$$

where $\{x(k)\}$ is a dynamics driven by $\{A(k)\}$.

If we furthermore assume that $\{A(k)\}$ is a doubly stochastic chain, then $\{\frac{1}{m}e\}$ is an absolute probability sequence for $\{A(k)\}$. Therefore, in this case we have $p^* = \frac{1}{m}$ and hence,

$$V(x((q+1)B)) \leq \left(1 - \frac{\gamma}{2m(m-1)}\right) V(x(qB)) \leq \left(1 - \frac{\gamma}{2m^2}\right) V(x(qB)), \tag{5.17}$$

where $V(x) = \frac{1}{m} \sum_{i=1}^{m} (x_i - \frac{1}{m}e^T x)^2$. This result is the same as the fast rate of convergence given in Theorem 10 in [4]. On the other hand, for a general B-connected chain, we have an absolute probability sequence $\{\pi(k)\} \geq \gamma^{(m-1)B}$ as shown below.

Lemma 5.8 *Let $\{A(k)\}$ be a B-connected chain. Then, $\{A(k)\}$ admits an absolute probability sequence such that $\{\pi(k)\} \geq \gamma^{(m-1)B}$.* **Q.E.D.**

Proof By Theorem 2.4, $\{A(k)\}$ is an ergodic chain with $\lim_{k \to \infty} A(k:t_0) = e\pi^T(t_0)$ with $\pi(t_0) \geq \gamma^{(m-1)B}$ for any $t_0 \geq 0$. Thus by Lemma 4.4, it follows that $\{\pi(k)\}$ is an absolute probability sequence for $\{A(k)\}$ and $\{\pi(k)\} \geq \gamma^{(m-1)B}$. **Q.E.D.**

Thus, by Lemma 5.8 and Corollary 5.4, it follows that

$$
V_\pi(x((q+1)B), (q+1)B) \leq \left(1 - \frac{\gamma^{(m-1)B+1}}{2(m-1)}\right)^q V_\pi(x(0), 0), \qquad (5.18)
$$

which is similar to the slow upper bound for the rate of convergence for general B-connected chains, as discussed in Theorem 8.1 in [5], that is often derived using the Lyapunov function $d(x) = \max_{i \in [m]} x_i - \min_{j \in [m]} x_j$.

Therefore, Theorem 5.5, results in different rate of convergence results for averaging dynamics. Comparing the slow rate of convergence result for general B-connected chains in (5.18) and the fast rate of convergence result for doubly stochastic B-connected chains in (5.17), we can see that the lower bound p^* on the entries of an absolute probability sequence $\{\pi(k)\}$ plays an important role in the rate of convergence of averaging dynamics.

5.4 Link Failure Models

Consider a random averaging scheme driven by an independent random chain $\{W(k)\}$. Suppose that an adversary randomly sets each of $W_{ij}(k)$ to zero. We refer to the resulted process as a link-failure process. In this section, we consider the effect of the link failure process on ergodicity and limiting behavior of an independent random chain $\{W(k)\}$.

Here, we assume that we have an underlying random chain and that there is another random process that models link failure in the random chain. We use $\{W(k)\}$ to denote the underlying adapted random chain, as in Eq. (3.1). We let $\{F(k)\}$ denote a link failure process, which is independent of the underlying chain $\{W(k)\}$. Basically, the failure process *reduces the information flow between agents* in the underlying random chain $\{W(k)\}$. For the failure process, we have either $F_{ij}(k) = 0$ or $F_{ij}(k) = 1$ for all $i, j \in [m]$ and $k \geq 0$, so that $\{F(k)\}$ is a binary matrix sequence. We define the *link-failure chain* as the random chain $\{U(k)\}$ given by

$$
U(k) = W(k) \cdot (ee^T - F(k)) + \text{diag}([W(k) \cdot F(k)]e), \qquad (5.19)
$$

where "$\cdot$" denotes the element-wise product of two matrices. To illustrate this chain, suppose that we have a random chain $\{W(k)\}$ and suppose that each entry $W_{ij}(k)$ is set to zero (fails), when $F_{ij}(k) = 1$. In this way, $F(k)$ induces a failure pattern on $W(k)$. The term $W(k) \cdot (ee^T - F(k))$ in Eq. (5.19) reflects this effect. Thus, $W(k) \cdot$

$(ee^T - F(k))$ does not have some of the entries of $W(k)$. This lack is compensated by the feedback term which is equal to the sum of the failed links, the term $\mathrm{diag}([W(k) \cdot F(k)]e)$. This is the same as adding $\sum_{j \neq i}[W(k) \cdot F(k)]_{ij}$ to the self-feedback weight $W_{ii}(k)$ of agent i at time k in order to ensure the stochasticity of $U(k)$.

Our discussion will be focused on a special class of link failure processes, which are introduced in the following definition.

Definition 5.4 A uniform link-failure process is a process $\{F(k)\}$ such that:

(a) The random variables $\{F_{ij}(k) \mid i, j \in [m], \ i \neq j\}$ are binary i.i.d. for any fixed $k \geq 0$.

(b) The process $\{F(k)\}$ is an independent process in time.

Note that the i.i.d. condition in Definition 5.4 is assumed for a fixed time. Therefore, the uniform link-failure chain can have a time-dependent distribution but for any given time the distribution of the link-failure should be identical across the different edges.

For the uniform-link failure process, we have the following result.

Lemma 5.9 *Let $\{W(k)\}$ be an independent random chain that is balanced and has feedback property. Let $\{F(k)\}$ be a uniform-link failure process that is independent of $\{W(k)\}$. Then, the failure chain $\{U(k)\}$ is infinite flow stable. Moreover, the link-failure chain is ergodic if and only if $\sum_{k=0}^{\infty}(1-p_k)\mathsf{E}[W_S(k)] = \infty$ for any non-trivial $S \subset [m]$, where $p_k = \mathrm{Pr}(F_{ij}(k) = 1)$.*

Proof By the definition of $\{U(k)\}$ in (5.19), the failure chain $\{U(k)\}$ is also independent since both $\{W(k)\}$ and $\{F(k)\}$ are independent. Then, for $i \neq j$ and for any $k \geq 0$, we have

$$\mathsf{E}\big[U_{ij}(k)\big] = \mathsf{E}\big[W_{ij}(k)(1 - F_{ij}(k))\big] = (1 - p_k)\bar{W}_{ij}(k), \qquad (5.20)$$

where the last equality holds since $W_{ij}(k)$ and $F_{ij}(k)$ are independent, and $\mathsf{E}\big[F_{ij}(k)\big] = p_k$. By summing both sides of relation (5.20) over $i \in S$ and $j \in \bar{S}$ and using the balanced property of $\{W(k)\}$, we obtain

$$\bar{U}_{S\bar{S}}(k) = (1 - p_k)\bar{W}_{S\bar{S}}(k) \geq \alpha(1 - p_k)\bar{W}_{\bar{S}S}(k) = \alpha\bar{U}_{\bar{S}S}(k),$$

where α is the balance-ness coefficient of $\{W(k)\}$. Thus, the failure chain $\{U(k)\}$ is a balanced chain.

We next show that $U(k)$ has feedback property. By the definition of $U(k)$, $U_{ii}(k) \geq W_{ii}(k)$ for all $i \in [m]$ and $k \geq 0$. Hence, $\mathsf{E}\big[U_{ii}(k)U_{ij}(k)\big] \geq \mathsf{E}\big[W_{ii}(k)U_{ij}(k)\big]$. Since $\{F(k)\}$ and $\{W(k)\}$ are independent, we have

$$\mathsf{E}\big[W_{ii}(k)U_{ij}(k)\big] = \mathsf{E}\big[\mathsf{E}[W_{ii}(k)U_{ij}(k) \mid F_{ij}(k)]\big] = \mathsf{E}\big[\mathsf{E}[W_{ii}(k)W_{ij}(k)(1 - F_{ij}(k)) \mid F_{ij}(k)]\big]$$
$$= (1 - p_k)\mathsf{E}\big[W_{ii}(k)W_{ij}(k)\big].$$

Thus, by the feedback property of $\{W(k)\}$, we have

$$\mathsf{E}\big[U_{ii}(k)U_{ij}(k)\big] \geq (1 - p_k)\gamma\mathsf{E}\big[W_{ij}(k)\big] = \gamma\mathsf{E}\big[U_{ij}(k)\big],$$

where the last equality follows from Eq. (5.20), and $\gamma > 0$ is the feedback coefficient for $\{W(k)\}$. Thus, $\{U(k)\}$ has feedback property with the same constant γ as the chain $\{W(k)\}$. Hence, the chain $\{U(k)\}$ satisfies the assumptions of Theorem 4.10, so the chain $\{U(k)\}$ is infinite flow stable. Therefore, it is ergodic if and only if $\sum_{k=0}^{\infty} \mathsf{E}[U_S(k)] = \infty$ for any nontrivial $S \subset [m]$. By Eq. (5.20) we have $\mathsf{E}[U_S(k)] = (1 - p_k)\mathsf{E}[W_S(k)]$, implying that $\{U(k)\}$ is ergodic if and only if $\sum_{k=0}^{\infty}(1 - p_k)\mathsf{E}[W_S(k)] = \infty$ for any nontrivial $S \subset [m]$. **Q.E.D.**

Lemma 5.9 shows that the severity of a uniform link failure process cannot cause instability in the system. An interesting feature of Lemma 5.9 is that if in the limit p_k are bounded away from 1 uniformly, i.e., $\limsup_{k \to \infty} p_k \leq \bar{p}$ for some $\bar{p} < 1$, then it can be seen that for any $i, j \in [m]$, $\sum_{k=0}^{\infty}(1 - p_k)\mathsf{E}[W_{ij}(k)] = \infty$ if and only if $\sum_{k=0}^{\infty} \mathsf{E}[W_{ij}(k)] = \infty$. Therefore, an edge $\{i, j\}$ belongs to the infinite flow graph of $\{W(k)\}$ if and only if it belongs to the infinite flow graph of $\{U(k)\}$. Hence, in this case, by Lemma 5.9 the following result is valid.

Corollary 5.5 *Let $\{W(k)\}$ be an independent random chain that is balanced and has feedback property. Let $\{F(k)\}$ be a uniform-link failure process that is independent of $\{W(k)\}$. For any $k \geq 0$, let $p_k = \Pr(F_{ij}(k) = 1)$ and suppose that $\limsup_{k \to \infty} p_k < 1$. Then, the ergodic behavior of the failure chain and the underlying chain $\{W(k)\}$ are the same.*

5.5 Hegselmann-Krause Model for Opinion Dynamics

In this section, we perform stability analysis of Hegselmann-Krause model [6]. Using the presented quadratic comparison function and some combinatorial arguments, we derive several bounds on the termination time for the Hegsemlann-Krause dynamics. In particular, we show that the Hegselmann-Krause dynamics terminates in at most $32m^4$ steps which results in a factor of m improvement of the previously known (upper) bound of $O(m^5)$.

5.5.1 Hegselmann-Krause Model

Suppose that we have a set of m agents and each of them has an opinion at time k, which is represented by a scalar $x_i(k) \in \mathbb{R}$. The vector $x(k) = (x_1(k), \ldots, x_m(k))^T \in \mathbb{R}^m$ of agent opinions is referred to as the *opinion profile* at time k. Starting from an initial profile $x(0) \in \mathbb{R}^m$, the opinion profile evolves in time as follows. Given a scalar $\epsilon > 0$, which we refer to as the *averaging radius*, at each time instance $k \geq 0$, agents whose opinions are within ϵ-difference will average their opinions. Formally, agent i shares its opinion with agents j in the set

$$\mathcal{N}_i(k) = \{j \in [m] \mid |x_i(k) - x_j(k)| \leq \epsilon\},$$

which contains i. The opinion of agent i at time $k + 1$ is the average of the opinions $x_j(k)$ for $j \in \mathcal{N}_i(k)$: at time k, i.e.,

$$x_i(k + 1) = \frac{1}{|\mathcal{N}_i(k)|} \sum_{j \in \mathcal{N}_i(k)} x_j(k) = B(k)x(k), \tag{5.21}$$

where $B_{ij}(k) = \frac{1}{|\mathcal{N}_i(k)|}$ for all $i \in [m]$ and $j \in \mathcal{N}_i(k)$, and $B_{ij}(k) = 0$ otherwise. Note that for a given $\epsilon > 0$ and an initial opinion profile $x(0) \in \mathbb{R}^m$, the dynamics $\{x(k)\}$ and the chain $\{B(k)\}$ are uniquely determined. We refer to $\{B(k)\}$ as the *chain generated by the initial profile $x(0)$*.

The asymptotic stability of the dynamics (5.21) has been shown in [7–9]. The asymptotic stability of the dynamics (5.21) can also be deduced from our developed results. To see this, note that we have $B_{ii}(k) \geq \frac{1}{m}$ for all i and $B_{S\bar{S}}(k) \geq \frac{1}{m}B_{\bar{S}S}(k)$ for all nontrivial $S \subset [m]$, implying that the chain $\{B(k)\}$ is balanced and has feedback property. Furthermore, the positive entries of $B(k)$ are bounded from below by $\gamma = \frac{1}{m}$. Thus by Lemma 4.12, we conclude that $\{B(k)\}$ has an absolute probability sequence $\{\pi(k)\}$ that is uniformly bounded by some $\mathcal{P}^*$ satisfying $\mathcal{P}^* \geq \frac{1}{m^{m-1}}$. As a result of Theorem 4.10, $\{B(k)\}$ is *infinite flow stable* and hence, the dynamics $\{x(k)\}$ is convergent.

5.5.2 Loose Bound for Termination Time of Hegselmann-Krause Dynamics

Here, we provide a loose bound for the convergence time of the Hegselmann-Krause dynamics which relies on the lower bound of an absolute probability sequence for the chains generated by Hegselmann-Krause dynamics.

Let us say that K is the *termination time*termination time for the Hegselmann-Krause dynamics, if $K \geq 0$ is the first time such that $x(k) = x(k + 1)$ for any $k > K$. Then, we have the following combinatorial result.

Lemma 5.10 *Suppose that $\{x(k)\}$ is the Hegselmann-Krause dynamics generated by an initial opinion profile $x(0) \in \mathbb{R}^m$. Suppose that $K \geq 0$ is such that $|x_\ell(K) - x_j(K)| \leq \frac{\epsilon}{2}$ for all $j, \ell \in \mathcal{N}_i(K)$ and all $i \in [m]$. Then $K + 1$ is a termination time for the dynamics $\{x(k)\}$.*

Proof First, we show that at time K, either $\mathcal{N}_i(K) = \mathcal{N}_j(K)$ or $\mathcal{N}_i(K) \bigcap \mathcal{N}_j(K) = \emptyset$ for any $i, j \in [m]$. To prove this, we argue by contraposition. So assume that $\mathcal{N}_i(K) \bigcap \mathcal{N}_j(K) \neq \emptyset$ and, without loss of generality (due to the symmetry), assume that there exists $\ell' \in \mathcal{N}_j(K) \setminus \mathcal{N}_i(K)$. Then, $|x_{\ell'}(K) - x_j(K)| \leq \frac{\epsilon}{2}$. Now, let $\ell \in \mathcal{N}_i(K) \bigcap \mathcal{N}_j(K)$ be arbitrary. We have $|x_\ell(K) - x_i(K)| \leq \frac{\epsilon}{2}$ and $|x_\ell(K) - x_j(K)| \leq \frac{\epsilon}{2}$, which by the triangle inequality implies $|x_i(K) - x_j(K)| \leq \epsilon$, thus showing that $i \in \mathcal{N}_j(K)$. By the definition of the time K and $i, \ell' \in \mathcal{N}_j(K)$, it follows that $|x_{\ell'}(K) - x_i(K)| \leq \frac{\epsilon}{2}$ implying $\ell' \in \mathcal{N}_i(K)$. This, however, contradicts the assumption $\ell' \in \mathcal{N}_j(K) \setminus \mathcal{N}_i(K)$.

Therefore, either $\mathcal{N}_i(K) = \mathcal{N}_j(K)$ or $\mathcal{N}_i(K) \bigcap \mathcal{N}_j(K) = \emptyset$. Thus, for the dynamics (5.21), we have

$$x_i(K+1) = x_j(K+1) = \frac{1}{|\mathcal{N}_i(K)|} \sum_{\ell \in \mathcal{N}_i(K)} x_\ell(K) \quad \text{for all } j \in \mathcal{N}_i(K) \text{ and all } i \in [m].$$

This implies $x_i(K+1) - x_j(K+1) = 0$ for all $j \in \mathcal{N}_i(K)$ and $i \in [m]$. Further, note that $|x_j(K+1) - x_\ell(K+1)| > \epsilon$ for all j, ℓ with $\mathcal{N}_j(K) \bigcap \mathcal{N}_\ell(K) = \emptyset$. Therefore, at time $K+1$, we have either $x_i(K+1) - x_j(K+1) = 0$ or $|x_i(K+1) - x_j(K+1)| > \epsilon$. Note that any such a vector is an equilibrium point of dynamics (5.21). Therefore, $x(k) = x(k+1)$ for all $k > K$. **Q.E.D.**

Based on Lemma 5.10, we have the following result for the termination time of the Hegselmann-Krause model.

Theorem 5.6 *Let $x(0) \in \mathbb{R}^m$ and let $\{x(k)\}$ be the corresponding dynamics driven by Hegselmann-Krause model for some averaging radius $\epsilon > 0$. Let $\{B(k)\}$ be the chain generated by $x(0)$. Then, $4m^2 \frac{d^2(x(0))}{p^* \epsilon^2}$ is a termination time for the dynamics, where $p^* > 0$ satisfies $\{\pi(k)\} \geq p^*$ for an absolute probability sequence $\{\pi(k)\}$ of the chain $\{B(k)\}$.*

Proof Let $K \geq 0$ be the first time that $|x_\ell(K) - x_j(K)| \leq \frac{\epsilon}{2}$ for all $j, \ell \in \mathcal{N}_i(K)$ and all $i \in [m]$.

By Corollary 4.3, we have

$$d^2(x(0)) \geq V_\pi(x(0), 0) \geq \sum_{k=0}^{K-1} \sum_{i<j} H_{ij}(k)(x_i(k) - x_j(k))^2$$

$$\geq p^* \sum_{k=0}^{K-1} \sum_{i<j} M_{ij}(k)(x_i(k) - x_j(k))^2,$$

where $H(k) = B(k)^T \operatorname{diag}(\pi(k+1))B(k)$, $M(k) = B(k)^T B(k)$ and the last inequality follows from $H_{ij}(k) \geq p^* M_{ij}(k)$ for all $i, j \in [m]$ and any $k \geq 0$ (by $\pi_\ell(k) \geq p^*$).

By the definition of time K, for $k < K$, there exist $i \in [m]$ and $j, \ell \in \mathcal{N}_i(k)$ such that $|x_\ell(k) - x_j(k)| > \frac{\epsilon}{2}$. But

$$M^{jT}(k)M^\ell(k) \geq B_{ij}(k)B_{i\ell}(k) \geq \frac{1}{m^2},$$

which follows from $j, \ell \in \mathcal{N}_i(k)$. Therefore, for $k < K$, we have:

$$\sum_{i<j} M_{ij}(k)(x_i(k) - x_j(k))^2 \geq \frac{1}{m^2} \frac{\epsilon^2}{4}.$$

Hence, it follows

$$p^* \frac{K\epsilon^2}{4m^2} \le d^2(x(0)),$$

implying $K \le 4m^2 \frac{d^2(x(0))}{p^*\epsilon^2}$. Therefore, by Lemma 5.10, $K = 4m^2 \frac{d^2(x(0))}{p^*\epsilon^2}$ is a termination time for the Hegselmann-Krause dynamics. **Q.E.D.**

Now, consider an initial profile $x(0) \in \mathbb{R}^m$ and without loss of generality assume that $x_1(0) \le x_2(0) \le \cdots \le x_m(0)$. Then if $x_{i+1}(0) - x_i(0) \le \epsilon$ for all $i \in [m-1]$, then we have $d(x(0)) = x_m(0) - x_1(0) \le m\epsilon$. Therefore, in this case the termination time would be less than or equal to

$$4m^2 \frac{d^2(x(0))}{p^*\epsilon^2} \le 4 \frac{m^4}{p^*}.$$

Note that if $x_{i+1}(0) - x_i(0) > \epsilon$ for some $i \in [m-1]$, then based on the form of the dynamics, we have $x_{i+1}(k) - x_i(k) > \epsilon$ for any $k \ge 0$. Therefore, in this case, we have a dynamics operating on each connected component of the initial profile. Hence, the termination time would be no larger than the termination time of the largest connected component which is less than $4\frac{m^4}{p^*}$. Therefore, the following result holds.

Corollary 5.6 *Let $x(0) \in \mathbb{R}^m$ and let $\{x(k)\}$ be the corresponding dynamics driven by Hegselmann-Krause model for some $\epsilon > 0$ and let $\{B(k)\}$ be the chain generated by the initial profile $x(0) \in \mathbb{R}^m$. Then, $4\frac{m^4}{p^*}$ is an upper bound for the termination time of the dynamics, where p^* satisfies $\{\pi(k)\} \ge p^*$ for an absolute probability sequence $\{\pi(k)\}$ for $\{B(k)\}$.*

Note that the provided bound in Corollary 5.6 does not depend on the averaging radius $\epsilon > 0$, as well as the initial opinion profile $x(0)$ and its spread $d(x(0))$.

5.5.3 An Improved Bound for Termination Time of Hegselmann-Krause Model

In Theorem 4.15 in [10], an upper bound of $O(m^5)$ is given for the termination time of the Hegselmann-Krause dynamics. Here, using the decreasing estimate of a quadratic comparison function that is provided in Eq. (4.8), we prove an $O(m^4)$ bound for the termination time of the Hegselmann-Krause dynamics.

Let $\{x(k)\}$ be the Hegselmann-Krause dynamics started at an initial opinion profile $x(0) \in \mathbb{R}^m$ with an averaging radius $\epsilon > 0$ and let $\{B(k)\}$ be the stochastic chain generated by $x(0) \in \mathbb{R}^m$. Throughout the subsequent discussion, without loss of generality, we assume that $\{x(k)\}$ is ordered, i.e. $x_1(k) \le x_2(k) \le \cdots \le x_m(k)$ which is allowed by the order preserving property of the Hegselmann-Krause dynamics.

By the preceding discussion, the dynamics $\{x(k)\}$ converges in finite time K. Also, for $k > K$, we can partition $[m]$ into $T_1, \ldots, T_p$ subsets such that $x_i(k) = x_j(k)$ for any $i, j \in T_r$ and for any $r \in \{1, \ldots, p\}$. Moreover, $|x_i(k) - x_j(k)| > \epsilon$ for all i, j belonging to different subsets T_r. Therefore, we have:

$$
B(k) = \begin{bmatrix} \frac{1}{|T_1|}J_{|T_1|} & \mathbf{0} & \cdots & \mathbf{0} & \mathbf{0} \\ \mathbf{0} & \frac{1}{|T_2|}J_{|T_2|} & \cdots & \mathbf{0} & \mathbf{0} \\ \vdots & \vdots & \ddots & \vdots & \vdots \\ \mathbf{0} & \mathbf{0} & \cdots & \frac{1}{|T_{p-1}|}J_{|T_{p-1}|} & \mathbf{0} \\ \mathbf{0} & \mathbf{0} & \cdots & \mathbf{0} & \frac{1}{|T_p|}J_{|T_p|} \end{bmatrix} \quad \text{for } k \geq K, \quad (5.22)
$$

where J_R is an $R \times R$ matrix with all entries equal to one, and the $\mathbf{0}$ s are zero matrices of appropriate dimensions. We refer to each T_r as a final component.final component

For a final component T_r and any $k \geq K$, let $\pi(k) = \frac{1}{|T_r|}\sum_{i \in T_r} e_i$. Then, by the form of $B(k)$ in (5.22), we have for any $k \geq K$,

$$
\pi^T(k) = \pi^T(k+1)B(k).
$$

For $k < K$, recursively, let $\pi^T(k) = \pi^T(k+1)B^T(k)$. Then, $\{\pi(k)\}$ is an absolute probability sequence for $\{B(k)\}$. We refer to such an absolute probability sequence as the absolute probability sequence started at the final component T_r, or simply, an absolute probability sequence started at a final component.

We say that the time K is the termination time for the final component T_r termination time| for a final component if $x_i(k) = x_j(k)$ for any $i, j \in T_r$ for any $k \geq K$ and K is the smallest number with this property.

For an agent $i \in [m]$, let the upper neighbor of i at time $k \geq 0$ be

$$
\bar{i}(k) = \min\{\ell \in \mathcal{N}_i(k) \mid x_i(k) < x_\ell(k)\}.
$$

Similarly, let the lower neighbor of $i \in [m]$ at time $k \geq 0$ be

$$
\underline{i}(k) = \max\{\ell \in \mathcal{N}_i(k) \mid x_\ell(k) < x_i(k)\}.
$$

An illustration of an agent i and its upper and lower neighbors is provided in Fig. 5.2. Note that the upper neighbor of $i \in [m]$ may not exist. This happens if $\{j \in [m] \mid x_j(k) \in (x_i(k), x_i(k) + \epsilon]\} = \emptyset$. Similarly, the lower neighbor may not exist.

To establish the convergence rate result for the Hegselmann-Krause dynamics, we make use of a sequence of intermediate results.

Lemma 5.11 *Let $\{x(k)\}$ be the Hegselmann-Krause dynamics started at an ordered initial opinion profile $x(0) \in \mathbb{R}^m$. Let $\{\pi(k)\}$ be an absolute probability sequence for $\{x(k)\}$ started at a final component T_r. Suppose that $k < K - 1$, where K is the terminating time for T_r. Then, for any $i \in [m]$ with $\pi_i(k+1) > 0$, there exists $\tau \in \mathcal{N}_i(k)$ such that $\pi_\tau(k+1) \geq \frac{1}{2}\pi_i(k+1)$ and $\mathcal{N}_\tau(k) \neq \mathcal{N}_i(k)$.*

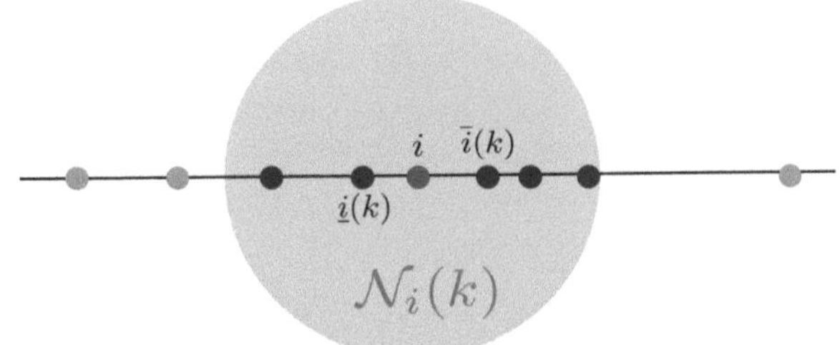

Fig. 5.2 Illustration of an agent i, its upper neighbor $\bar{i}(k)$ and its lower neighbor $\underline{i}(k)$

Proof Suppose that $k < K - 1$. Let $i \in [m]$ and let $\mathcal{N}_i^-(k+1)$ and $\mathcal{N}_i^+(k+1)$ be the subsets of $\mathcal{N}_i(k+1)$ defined by:

$$\mathcal{N}_i^-(k+1) = \{j \in \mathcal{N}_i(k+1) \mid x_j(k+1) \leq x_i(k+1)\},$$

and

$$\mathcal{N}_i^+(k+1) = \{j \in \mathcal{N}_i(k+1) \mid x_j(k+1) \geq x_i(k+1)\}.$$

By the definition of an absolute probability sequence, we have

$$\pi_i(k+1) = \sum_{j \in \mathcal{N}_i(k+1)} \pi_j(k+2) B_{ji}(k+1)$$

$$\leq \sum_{j \in \mathcal{N}_i^-(k+1)} \pi_j(k+2) B_{ji}(k+1) + \sum_{j \in \mathcal{N}_i^+(k+1)} \pi_j(k+2) B_{ji}(k+1),$$

$$(5.23)$$

where the inequality follows by the fact that $\mathcal{N}_i^-(k+1) \bigcup \mathcal{N}_i^+(k+1) = \mathcal{N}_i(k+1)$. Thus, either of the following inequalities holds:

$$\frac{\pi_i(k+1)}{2} \leq \sum_{j \in \mathcal{N}_i^-(k+1)} \pi_j(k+2) B_{ji}(k+1)$$

$$\text{or} \quad \frac{\pi_i(k+1)}{2} \leq \sum_{j \in \mathcal{N}_i^+(k+1)} \pi_j(k+2) B_{ji}(k+1).$$

Without loss of generality assume that $\frac{\pi_i(k+1)}{2} \leq \sum_{j \in \mathcal{N}_i^+(k+1)} \pi_j(k+2) B_{ji}(k+1)$.
Now, consider the following cases:

(1) $\bar{i}(k+1)$ exists: then we have $\bar{i}(k+1) \in \mathcal{N}_i^+(k+1)$ and $\mathcal{N}_i^+(k+1) \subseteq \mathcal{N}_{\bar{i}(k+1)}(k+1)$. Thus, we have:

$$\pi_{\bar{i}(k+1)}(k+1) = \sum_{j \in \mathcal{N}_{\bar{i}}(k+1)} \pi_j(k+2) B_{j\bar{i}}(k+1) \geq \sum_{j \in \mathcal{N}_i^+(k+1)} \pi_j(k+2) B_{j\bar{i}}(k+1)$$

$$= \sum_{j \in \mathcal{N}_i^+(k+1)} \pi_j(k+2) B_{ji}(k+1) \geq \frac{\pi_i(k+1)}{2},$$

where, in the second inequality, we used the fact that the positive entries in each row of $B(k+1)$ are identical. Thus, $\pi_{\bar{i}(k+1)}(k+1) \geq \frac{\pi_i(k+1)}{2}$. Note that $\bar{i}(k+1) \in \mathcal{N}_i(k)$, because otherwise,

$$x_{\bar{i}(k+1)}(k) \leq x_{\bar{i}(k+1)}(k+1),$$

and also,

$$x_i(k+1) \leq x_i(k).$$

Since $\bar{i}(k+1) \in \mathcal{N}_i^+(k+1)$, it follows $\bar{x}_i(k+1) - x_i(k+1) \leq \epsilon$. This and the preceding two relations imply $x_{\bar{i}(k+1)}(k) - x_i(k) \leq \epsilon$, i.e. $\bar{i}(k+1) \in \mathcal{N}_i(k)$. Also, $\mathcal{N}_{\bar{i}(k+1)}(k) \neq \mathcal{N}_i(k)$, because otherwise, $x_i(k+1) = x_{\bar{i}(k+1)}(k+1)$ which contradicts with $x_i(k+1) < x_{\bar{i}(k+1)}(k+1)$. Therefore, in this case, the assertion is true and we have $\tau = \bar{i}(k+1)$.

(2) $\bar{i}(k+1)$ does not exist: in this case, for any $j \in \mathcal{N}_i^+(k+1)$, we have $x_j(k+1) = x_i(k+1)$. Thus, $\mathcal{N}_i^+(k+1) \subseteq \mathcal{N}_i^-(k+1)$ and hence, $\mathcal{N}_i^-(k+1) = \mathcal{N}_i(k+1)$. If $\underline{i}(k+1)$ does not exist, then we have $\mathcal{N}_i^-(k+1) = \mathcal{N}_i(k+1)$, implying $x_j(k+1) = x_i(k+1)$ for any $j \in \mathcal{N}_i(k)$. This implies that $k+1$ is the termination time for the final component T_r which contradicts with the assumption $k < K - 1$.

Thus, in this case $\underline{i}(k+1)$ must exist. But since $\mathcal{N}_i(k) = \mathcal{N}_i^-(k+1) \subseteq \mathcal{N}_{\underline{i}(k+1)}(k+1)$, using the same line of argument as in the previous case, it follows that

$$\pi_{\underline{i}(k+1)} \geq \frac{1}{2}\pi_i(k+1),$$

and the assertion holds for $\tau = \underline{i}(k+1)$. **Q.E.D.**

For an agent $i \in [m]$, let $d_i(k) = \max_{j \in \mathcal{N}_i(k)} x_j(k) - \min_{j \in \mathcal{N}_i(k)} x_j(k)$. In a sense, $d_i(k)$ is the spread of the opinions that agent i observes at time k. Let us prove the following inequality.

Lemma 5.12 *Let $\{x(k)\}$ be the Hegselmann-Krause dynamics started at an ordered initial opinion profile $x(0) \in \mathbb{R}^m$. Then, for any $\ell \in [m]$, we have:*

$$\sum_{i,j \in \mathcal{N}_\ell(k)} (x_i(k) - x_j(k))^2 \geq \frac{1}{4}|\mathcal{N}_\ell(k)|d_\ell^2(k) \qquad \text{for any } k \geq 0.$$

Proof If $d_\ell(k) = 0$, then the assertion follows immediately. So, suppose that $d_\ell(k) \neq 0$. Let $lb = \min\{i \leq \ell \mid i \in \mathcal{N}_\ell(k)\}$ and $ub = \max\{i \geq \ell \mid i \in \mathcal{N}_\ell(k)\}$. In words, lb and ub are the agents with the smallest and largest opinion in the neighborhood of ℓ. Therefore, $d_\ell(k) = x_{ub}(k) - x_{lb}(k)$ and since $d_\ell(k) \neq 0$, we have $lb \neq ub$. Thus, we have

$$\sum_{i,j\in\mathcal{N}_\ell(k)} (x_i(k) - x_j(k))^2 \geq \sum_{j\in\mathcal{N}_\ell(k)} (x_{ub}(k) - x_j(k))^2 + \sum_{j\in\mathcal{N}_\ell(k)} (x_j(k) - x_{lb}(k))^2$$

$$= \sum_{j\in\mathcal{N}_\ell(k)} \left\{ (x_{ub}(k) - x_j(k))^2 + (x_j(k) - x_{lb}(k))^2 \right\}$$

$$\geq \sum_{j\in\mathcal{N}_\ell(k)} \frac{1}{4}(x_{ub}(k) - x_{lb}(k))^2 = \frac{1}{4}|\mathcal{N}_\ell(k)|d_\ell^2(k).$$

In the last inequality we used the fact that

$$(x_{ub}(k) - x_j(k))^2 + (x_j(k) - x_{lb}(k))^2 \geq \frac{1}{4}(x_{ub}(k) - x_{lb}(k))^2,$$

which holds since the function $s \to (x_{ub}(k)-s)^2+(s-x_{lb}(k))^2$ attains its minimum at $s = \frac{x_{lb}(k)+x_{ub}(k)}{2}$. **Q.E.D.**

Based on Lemma 5.12, we can prove another intermediate result which bounds the decrease value of the quadratic comparison function for the Hegselmann-Krause dynamics.

Lemma 5.13 Let $\{x(k)\}$ be the Hegselmann-Krause dynamics started at an ordered initial opinion profile $x(0) \in \mathbb{R}^m$. Then, for any $k \geq 0$, we have

$$\sum_{i<j}^m H_{ij}(k)(x_i(k) - x_j(k))^2 \geq \frac{1}{2}\sum_{\ell=1}^m \pi_\ell(k+1)\frac{d_\ell^2(k)}{4|\mathcal{N}_\ell(k)|} \geq \frac{1}{8m}\sum_{\ell=1}^m \pi_\ell(k+1)d_\ell^2(k),$$

where $H(k) = B^T(k)\mathrm{diag}(\pi(k+1))B(k)$, $\{B(k)\}$ is the stochastic chain generated by $x(0)$, and $\{\pi(k)\}$ is an absolute probability sequence for $\{B(k)\}$.

Proof Since $H(k)$ is a symmetric matrix, we have:

$$2\sum_{i<j}^m H_{ij}(k)(x_i(k) - x_j(k))^2 = \sum_{i=1}^m\sum_{j=1}^m H_{ij}(k)(x_i(k) - x_j(k))^2.$$

On the other hand, we have:

$$\sum_{i=1}^m\sum_{j=1}^m H_{ij}(k)(x_i(k) - x_j(k))^2 = \sum_{i=1}^m\sum_{j=1}^m\sum_{\ell=1}^m \pi_\ell(k+1)B_{\ell i}(k)B_{\ell j}(k)(x_i(k) - x_j(k))^2$$

$$= \sum_{\ell=1}^m \pi_\ell(k+1)\left(\sum_{i=1}^m\sum_{j=1}^m B_{\ell i}(k)B_{\ell j}(k)(x_i(k) - x_j(k))^2\right)$$

$$= \sum_{\ell=1}^m \frac{\pi_\ell(k+1)}{|\mathcal{N}_\ell(k)|^2} \sum_{i\in\mathcal{N}_\ell(k)}\sum_{j\in\mathcal{N}_\ell(k)} (x_i(k) - x_j(k))^2,$$

$$\tag{5.24}$$

where we used the fact that $B_{\ell r}(k) = \frac{1}{|\mathcal{N}_\ell(k)|}$ if and only if $r \in \mathcal{N}_\ell(k)$, and $B_{\ell r}(k) = 0$, otherwise. Thus, by Lemma 5.12 and relation (5.24), we have

$$\sum_{i=1}^{m}\sum_{j=1}^{m}H_{ij}(k)(x_i(k)-x_j(k))^2 \geq \sum_{\ell=1}^{m}\frac{\pi_\ell(k+1)}{4|\mathcal{N}_\ell(k)|^2}|\mathcal{N}_\ell(k)|d_\ell^2(k)$$

$$= \sum_{\ell=1}^{m}\pi_\ell(k+1)\frac{d_\ell^2(k)}{4|\mathcal{N}_\ell(k)|}.$$

Thus the first inequality follows. The second inequality follows from $|\mathcal{N}_\ell(k)| \leq m$. **Q.E.D.**

Now, we are ready to prove the $O(m^4)$ upper bound for the convergence time of the Hegselmann-Krause dynamics.

Theorem 5.7 *For the Hegselmann-Krause dynamics $\{x(k)\}$ started at an initial opinion profile $x(0) \in \mathbb{R}^m$, the termination time K is bounded above by $\frac{32m^2d^2(x(0))}{\epsilon^2} + 2$, where $\epsilon > 0$ is the averaging radius.*

Proof Let $\{B(k)\}$ be the stochastic chain generated by $x(0)$ and let $\{\pi(k)\}$ be an absolute probability sequence started at a final component T. Let $\bar{\ell} = \arg\max_{\ell\in[m]}\pi_\ell(k+1)$. Since, $\pi(k+1)$ is a stochastic vector, it follows that $\pi_{\bar{\ell}}(k+1) \geq \frac{1}{m}$. Now, suppose that $k < K-1$, where K is the termination time of T_r. Consider the following cases:

Case 1. $d_{\bar{\ell}}(k) > \frac{\epsilon}{2}$: in this case, by Lemma 5.13, we have

$$\sum_{i<j}H_{ij}(k)(x_i(k)-x_j(k))^2 \geq \frac{1}{8m}\sum_{\ell=1}^{m}\pi_\ell(k+1)d_\ell^2(k)$$

$$\geq \frac{1}{8m}\pi_{\bar{\ell}}(k+1)d_{\bar{\ell}}^2(k) \geq \frac{\epsilon^2}{32m^2},$$

which follows from $\pi_{\bar{\ell}}(k+1) \geq \frac{1}{m}$ and $d_{\bar{\ell}}(k) > \frac{\epsilon}{2}$.

Case 2. $d_{\bar{\ell}}(k) \leq \frac{\epsilon}{2}$: since $k < K-1$, by Lemma 5.11, there exists $\tau \in \mathcal{N}_{\bar{\ell}}(k)$ such that

$$\pi_\tau(k+1) \geq \frac{1}{2}\pi_{\bar{\ell}}(k+1) \geq \frac{1}{2m}.$$

Moreover, $\mathcal{N}_\tau(k) \neq \mathcal{N}_{\bar{\ell}}(k)$.

Let us first prove that $d_\tau(k) \geq \epsilon$. Since $d_{\bar{\ell}}(k) \leq \frac{\epsilon}{2}$, it follows that $|x_{\bar{\ell}}(k) - x_\tau(k)| \leq \frac{\epsilon}{2}$. Also, for any $i \in \mathcal{N}_{\bar{\ell}}(k)$, we have:

$$|x_i(k) - x_\tau(k)| \leq |x_i(k) - x_{\bar{\ell}}(k)| + |x_{\bar{\ell}}(k) - x_\tau(k)| \leq \epsilon,$$

implying $i \in \mathcal{N}_\tau(k)$. Thus, $\mathcal{N}_{\bar{\ell}}(k) \subset \mathcal{N}_\tau(k)$. Let $i \in \mathcal{N}_\tau(k) \setminus \mathcal{N}_{\bar{\ell}}(k)$. Then, $|x_i(k) - x_{\bar{\ell}}(k)| > \epsilon$, and hence, $d_\tau(k) \geq \epsilon$.
Therefore, by Lemma 5.11, we have:

$$\sum_{i<j} H_{ij}(k)(x_i(k)-x_j(k))^2 \geq \frac{1}{8m}\sum_{\ell=1}^{m}\pi_\ell(k+1)d_\ell^2(k) \geq \frac{1}{8m}\pi_\tau(k+1)d_\tau^2(k) \geq \frac{\epsilon^2}{16m^2},$$

which follows from $\pi_\tau(k+1) \geq \frac{1}{2m}$ and $d_\tau(k) \geq \epsilon$.

All in all, if $k < K - 1$, we have

$$\sum_{i<j} H_{ij}(k)(x_i(k) - x_j(k))^2 \geq \frac{\epsilon^2}{32m^2}.$$

At the same time, by Theorem 4.4, we should have:

$$d^2(x(0)) \geq V_\pi(x(0), 0) \geq \sum_{k=0}^{K-1}\sum_{i<j} H_{ij}(k)(x_i(k) - x_j(k))^2.$$

Thus, we should have $K \leq \frac{32m^2 d^2(x(0))}{\epsilon^2} + 2$. But the derived bound does not rely on the particular choice of the final component T and hence, $\frac{32m^2 d^2(x(0))}{\epsilon^2} + 2$ is a termination time for the Hegselmann-Krause dynamics. **Q.E.D.**

Note, that the proof of Theorem 5.7, does not rely on the existence of an absolute probability sequence $\{\pi(k)\}$ with $\{\pi(k)\} \geq p^*$ for some scalar $p^* > 0$. It only relies on the decrease estimate given in Theorem 4.4.

As discussed in Corollary 5.6, without loss of generality, we can assume that the graph induced by the initial opinion profile $x(0) \in \mathbb{R}^m$ is connected. Otherwise, we can use the given analysis in Theorem 5.7 for the largest connected component of the initial profile. When the graph is connected, we have $d(x(0)) \leq m\epsilon$. Thus, $\frac{32m^2 d^2(x(0))}{\epsilon^2} \leq 32m^4$ and hence, $32m^4 + 1$ is an upper bound for the termination time of the Hegselmann-Krause dynamics.

Corollary 5.7 $32m^4 + 1$ *is an upper bound for the termination time of the Hegselmann-Krause dynamics.*

5.6 Alternative Proof of Borel-Cantelli lemma

In this section, we provide an alternative proof for the second Borel-Cantelli lemma (Lemma A.6) based on the results developed so far.

For a sequence of events $\{E_k\}$ in some probability space, let us define a sequence of random 2×2 stochastic matrices $\{W(k)\}$ as follows:

$$W(k) = \frac{1}{2}ee^T \mathbf{1}_{E_k} + I\mathbf{1}_{E_k^c},$$

for $k \geq 0$, i.e., $W(k)$ is equal to the stochastic matrix $\frac{1}{2}ee^T$ on E_k and otherwise, it is equal to the identity matrix.

Let $U = \bigcap_{k=0}^{\infty} \bigcup_{s=k}^{\infty} E_s = \{E_k \; i.o.\}$ where *i.o.* stands for infinitely often. Then, $\omega \in U$ means that in the corresponding random chain $\{W(k)\}(\omega)$ we have $W(k)(\omega) = \frac{1}{2}e^T e$ for infinitely many indices k. Let $\{A(k)\}$ be a chain of 2×2 matrices where $A(k)$ is either I or $\frac{1}{2}ee^T$ for all $k \geq 0$. Since $M(\frac{1}{2}ee^T) = \frac{1}{2}ee^T$ for any 2×2 stochastic matrix M, it follows that the chain $\{A(k)\}$ is ergodic if and only if the matrix $\frac{1}{2}ee^T$ appears infinitely many times in the chain. This observation, together with Theorem 5.1, gives rise to an alternative proof of the second Borel-Cantelli lemma.

Theorem 5.8 (*Second Borel-Cantelli lemma*) *Let* $\{E_k\}$ *be independent and* $\sum_{k=0}^{\infty} \Pr(E_k) = \infty$. *Then* $\Pr(U) = 1$.

Proof Let $\{W(k)\}$ be the random chain corresponding to the sequence $\{E_k\}$. Then $\{W(k)\}(\omega)$ is ergodic if and only if $\omega \in U$. Since $\sum_{k=0}^{\infty} \Pr(E_k) = \infty$, it follows $\sum_{k=0}^{\infty} \mathsf{E}[W_{12}(k) + W_{21}(k)] = \infty$, implying that $\{W(k)\}$ has infinite flow. Furthermore, since $\{E_k\}$ is independent, so is $\{W(k)\}$. Observe that each realization of $W(k)$ is doubly stochastic. We also have $W_{ii}(k) \geq \frac{1}{2}$ for $i = 1, 2$ and any k. Therefore, by Theorem 5.1, the model $\{W(k)\}$ is ergodic and, hence, $\Pr(U) = 1$. **Q.E.D.**

In the derivation of the above proof, there is a possibility of being exposed to the trap of circular reasoning. But to the best of author's knowledge, none of the steps in the proof is involving with the use of Lemma A.6 itself.

References

1. A. Tahbaz-Salehi, A. Jadbabaie, A necessary and sufficient condition for consensus over random networks. IEEE Trans. Autom. Control **53**(3), 791–795 (2008)
2. P. Kumar, P. Varaiya, *Stochastic Systems Estimation Identification and Adaptive Control*, Information and System Sciences Series (Englewood Cliffs New Jersey, Prentice–Hall, 1986)
3. S. Meyn, R. Tweedie, *Markov Chains and Stochastic Stability*, Cambridge Mathematical Library (Cambridge University Press, Cambridge, 2009)
4. A. Nedić, A. Olshevsky, A. Ozdaglar, J. Tsitsiklis, On distributed averaging algorithms and quantization effects. IEEE Trans. Autom. Control **54**(11), 2506–2517 (2009)
5. A. Olshevsky, J. Tsitsiklis, Convergence speed in distributed consensus and averaging. SIAM J. Control Optim. **48**(1), 33–55 (2008)
6. R. Hegselmann, U. Krause, Opinion dynamics and bounded confidence models, analysis, and simulation. J. Artif. Soc. Social Simul. **5**, 3 (2002)
7. L. Moreau, Stability of multi-agent systems with time-dependent communication links. IEEE Trans. Autom. Control **50**(2), 169–182 (2005)
8. J. Lorenz, A stabilization theorem for continuous opinion dynamics. Physica A **355**, 217–223 (2005)
9. J.M. Hendrickx, Graphs and networks for the analysis of autonomous agent systems. Ph.D. dissertation, Université Catholique de Louvain, 2008
10. F. Bullo, J. Cortés, S. Martínez, *Distributed Control of Robotic Networks*, Applied Mathematics Series (Princeton University Press, Princeton, 2009) (electronically available at http:// coordinationbook.info)

Chapter 6
Absolute Infinite Flow Property

Motivated by the concept of the infinite flow property, in this chapter we introduce the concept of absolute infinite flow property and extend some of the results developed so far. Our discussion in this chapter is restricted to deterministic chains and deterministic dynamics.

In Sect. 6.1, we introduce the concepts of a regular sequence and flow over a regular sequence, as well as the absolute infinite flow property. Then, in Sect. 6.2 we prove that the absolute infinite flow property is necessary for ergodicity. We do this through the rotational transformation of a chain with respect to a permutation chain. In Sect. 6.3, we introduce the class of decomposable chains for which their absolute infinite flow property can be computed more efficiently as compared to a general chain. Finally in Sect. 6.4, we consider a subclass of decomposable chains, the doubly stochastic chains, and prove that the absolute infinite flow property is equivalent to ergodicity for those chains. We also prove that the product of any sequence of doubly stochastic matrices is essentially convergent, i.e., it is convergent up to a permutation sequence.

6.1 Absolute Infinite Flow

In this section, we introduce the absolute infinite flow property which will play a central role in the forthcoming discussion in this chapter. To introduce this property, let us provide a visualization scheme for the dynamics in Eq. (2.2) which is motivated by the *trellis diagram* method for visualizing convolution decoders. Let us introduce the *trellis graph* associated with a given stochastic chain. The trellis graph of a stochastic chain $\{A(k)\}$ is an infinite directed weighted graph $G = (V, \mathcal{E}, \{A(k)\})$, with the vertex set V equal to the infinite grid $[m] \times \mathbb{Z}^+$ and the edge set

$$\mathcal{E} = \{((j, k), (i, k + 1)) \mid j, i \in [m], \quad k \geq 0\}. \tag{6.1}$$

In other words, we consider a copy of the set $[m]$ for each time $k \geq 0$ and we stack these copies over time, thus generating the infinite vertex set $V = \{(i, k) \mid i \in [m], k \geq 0\}$.

B. Touri, *Product of Random Stochastic Matrices and Distributed Averaging*,
Springer Theses, DOI: 10.1007/978-3-642-28003-0_6,
© Springer-Verlag Berlin Heidelberg 2012

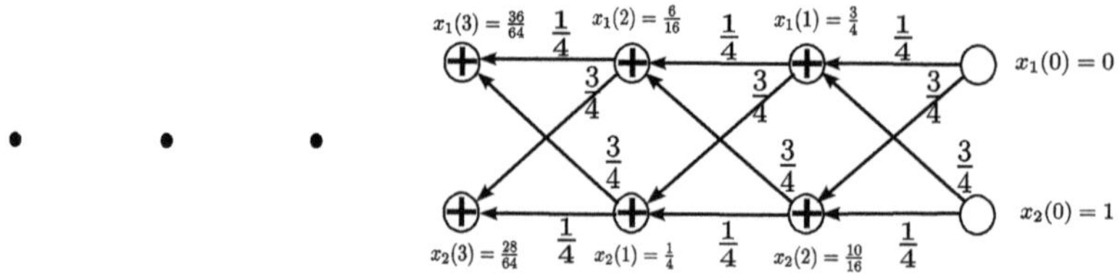

Fig. 6.1 The trellis graph of the 2×2 chain $\{A(k)\}$ with weights $A(k)$ given in Eq. (6.2)

We then place a link from each $j \in [m]$ at time k to every $i \in [m]$ at time $k+1$, i.e., a link from each vertex $(j, k) \in V$ to each vertex $(i, k + 1) \in V$. Finally, we assign the weight $A_{ij}(k)$ to the link $((j, k), (i, k + 1))$. Now, consider the whole graph as an information tunnel through which the information flows: we inject a scalar $x_i(0)$ at each vertex $(i, 0)$ of the graph. Then, from this point on, at each time $k \geq 0$, the information is transferred from time k to time $k+1$ through each edge of the graph that acts as a communication link. Each link attenuates the in-vertex's value with its weight, while each vertex sums the information received through the incoming links. One can observe that the resulting information evolution is the same as the dynamics given in Eq. (2.2). As a concrete example, consider the 2×2 static chain $\{A(k)\}$ with $A(k)$ defined by:

$$A(k) = \begin{bmatrix} \frac{1}{4} & \frac{3}{4} \\[2mm] \frac{3}{4} & \frac{1}{4} \end{bmatrix} \quad \text{for } k \geq 0. \tag{6.2}$$

The trellis graph of this chain is depicted in Fig. 6.1.

Recall the definition of infinite flow property for a chain $\{A(k)\}$ (Definition 3.2) which requires that $\sum_{k=0}^{\infty} A_S(k) = \infty$ for any non-trivial $S \subset [m]$. Graphically, the infinite flow property requires that in the trellis graph of a given model $\{A(k)\}$, the weights on the edges between $S \times \mathbb{Z}^+$ and $\bar{S} \times \mathbb{Z}^+$ add up to infinity, for any non-trivial $S \subset [m]$.

Although infinite flow property is necessary for ergodicity, this property alone is not *strong enough* to separate some stochastic chains from ergodic chains, such as permutation sequences. As a concrete example consider a static chain $\{A(k)\}$ of permutation matrices $A(k)$ given by

$$A(k) = \begin{bmatrix} 0 & 1 \\ 1 & 0 \end{bmatrix} \quad \text{for } k \geq 0. \tag{6.3}$$

The chain $\{A(k)\}$ has infinite flow property, but it is not ergodic.

As a remedy for this situation, in Chap. 4 , we have imposed additional conditions on the matrices $A(k)$ that eliminate permutation matrices such as feedback properties. Here, we take a different approach.

Specifically, we will require a stronger infinite flow property by letting the set S in Definition 3.2 vary with time. In order to do so, we will consider sequences

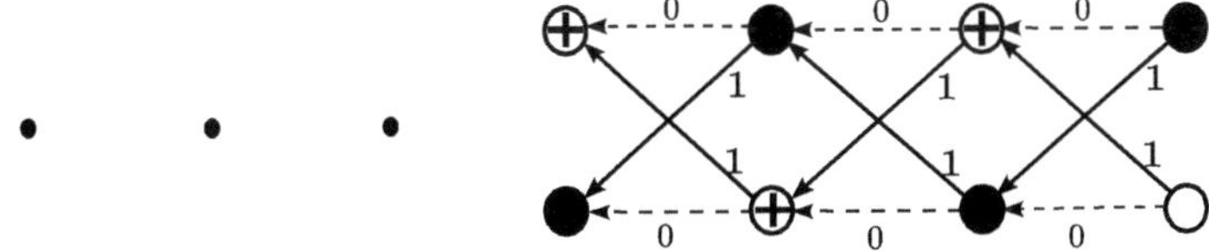

Fig. 6.2 The trellis graph of the permutation chain in Eq. (6.3). For the regular sequence $\{S(k)\}$, with $S(k) = \{1\}$ if k is even and $S(k) = \{2\}$ otherwise, the vertex set $\{(i, k) \mid i \in S(k), k \geq 0\}$ is marked by *black vertices*. The flow $F(\{A(k)\}; \{S(k)\})$, as defined in (6.5), corresponds to the summation of the weights on the *dashed edges*

$\{S(k)\}$ of non-trivial index sets $S(k) \subset [m]$ with a form of regularity in the sense that the sets $S(k)$ have the same cardinality for all k. In what follows, we will reserve notation $\{S(k)\}$ for the sequences of index sets $S(k) \subset [m]$. Furthermore, for easier exposition, we define the notion of regularity for $\{S(k)\}$ as follows.

Definition 6.1 A sequence $\{S(k)\}$ is regular if the sets $S(k)$ have the same (nonzero) cardinality, i.e., $|S(k)| = |S(0)|$ for all $k \geq 0$ and $|S(0)| \neq 0$.

The nonzero cardinality requirement in Definition 6.1 is imposed only to exclude the trivial sequence $\{S(k)\}$ consisting of empty sets.

Graphically, a regular sequence $\{S(k)\}$ corresponds to the subset $\{(i, k) \mid i \in S(k), k \geq 0\}$ of vertices in the trellis graph associated with a given chain. As an illustration, let us revisit the 2×2 chain given in Eq. (6.3). Consider the regular $\{S(k)\}$ defined by

$$S(k) = \begin{cases} \{1\} & \text{if } k \text{ is even,} \\ \{2\} & \text{if } k \text{ is odd.} \end{cases}$$

The vertex set $\{(i, k) \mid i \in S(k), k \geq 0\}$ associated with $\{S(k)\}$ is shown in Fig. 6.2.

Now, let us consider a chain $\{A(k)\}$ of stochastic matrices $A(k)$. Let $\{S(k)\}$ be any regular sequence. At any time k, we define the flow associated with the entries of the matrix $A(k)$ across the index sets $S(k + 1)$ and $S(k)$ as follows:

$$A_{S(k+1),S(k)}(k) = \sum_{i \in S(k+1), j \in \bar{S}(k)} A_{ij}(k) + \sum_{i \in \bar{S}(k+1), j \in S(k)} A_{ij}(k) \quad \text{for } k \geq 0. \tag{6.4}$$

The flow $A_{S(k+1),S(k)}(k)$ could be viewed as an instantaneous flow at time k induced by the corresponding elements in the matrix chain and the index set sequence. Accordingly, we define the total flow of a chain $\{A(k)\}$ over $\{S(k)\}$, as follows:

$$F(\{A(k)\}; \{S(k)\}) = \sum_{k=0}^{\infty} A_{S(k+1),S(k)}(k). \tag{6.5}$$

We are now ready to extend the definition of infinite flow property to time-varying index sets $S(k)$.

Definition 6.2 A stochastic chain $\{A(k)\}$ has absolute infinite flow property if

$$F(\{A(k)\}; \{S(k)\}) = \infty \quad \text{for every regular sequence} \{S(k)\}.$$

Note that the absolute infinite flow property of Definition 6.2 is more restrictive than the infinite flow property of Definition 3.2. In particular, we can see this by letting the set sequence $\{S(k)\}$ be static, i.e., $S(k) = S$ for all k and some nonempty $S \subset [m]$. In this case, the flow $A_{S(k+1),S(k)}(k)$ across the index sets $S(k + 1)$ and $S(k)$ as defined in Eq. (6.4) reduces to $A_S(k)$, while the flow $F(\{A(k)\}; \{S(k)\})$ in Eq. (6.5) reduces to $\sum_{k=0}^{\infty} A_S(k)$. This brings us to the quantities that define the infinite flow property (Definition 3.2). Thus, the infinite flow property requires that the flow across a trivial regular sequence $\{S\}$ is infinite for all nonempty $S \subset [m]$, which is evidently less restrictive requirement than that of Definition 6.2. In light of this, we see that if a stochastic chain $\{A(k)\}$ has absolute infinite property, then it has infinite flow property.

The distinction between absolute infinite flow property and infinite flow property is actually much deeper. Recall our example of the chain $\{A(k)\}$ in Eq. (6.3), which demonstrated that a permutation chain may posses infinite flow property. Now, for this chain, consider the regular sequence $\{S(k)\}$ with $S(2k) = \{1\}$ and $S(2k + 1) = \{2\}$ for $k \geq 0$. The trellis graph associated with $\{A(k)\}$ is shown in Fig. 6.2, where $\{S(k)\}$ is depicted by black vertices. The flow $F(\{A(k)\}; \{S(k)\})$ corresponds to the summation of the weights on the dashed edges in Fig. 6.2, which is equal to zero in this case. Thus, the chain $\{A(k)\}$ does not have absolute flow property.

In fact, while some chains of permutation matrices may have infinite flow property, it turns out that no chain of permutation matrices has absolute infinite flow property. In other words, absolute infinite flow property is strong enough to filter out the chains of permutation matrices in our search for necessary and sufficient conditions for ergodicity which is a significant distinction between absolute infinite flow property and infinite flow property. To formally establish this property, we turn our attention to an intimate connection between a regular index sequence and a permutation sequence that can be associated with the index sequence.

Specifically, an important feature of a regular sequence $\{S(k)\}$ is that it can be obtained as the image of the initial set $S(0)$ under a certain permutation sequence $\{P(k)\}$. To see this, note that we can always find a one-to-one matching between the indices in $S(k)$ and $S(k + 1)$ since $|S(k)| = |S(k + 1)|$. The complements $\bar{S}(k)$ and $\bar{S}(k + 1)$ of the sets $S(k)$ and $S(k + 1)$, respectively, also have the same cardinality, so there is a one-to-one matching between $\bar{S}(k)$ and $\bar{S}(k + 1)$ as well. Thus, we have a matching for the indices in $[m]$ which is one-to-one mapping between $S(k)$ to $S(k + 1)$, and also one-to-one mapping between $\bar{S}(k)$ and $\bar{S}(k + 1)$. Therefore, we can define an $m \times m$ matrix $P(k)$ as the incidence matrix corresponding to this matching, as follows: for every $j \in S(k)$ we let $P_{ij}(k) = 1$ if index j is matched with the index $i \in \bar{S}(k + 1)$ and $P_{ij}(k) = 0$ otherwise; similarly, for every $j \in \bar{S}(k)$ we let $P_{ij}(k) = 1$ if index j is matched with the index $i \in \bar{S}(k + 1)$ and $P_{ij}(k) = 0$ otherwise. The resulting matrix $P(k)$ is a permutation matrix and the set $S(k + 1)$ is the image of set $S(k)$ under the permutation matrix $P(k)$, i.e., $S(k+1) = P(k)(S(k))$.

Continuing in this way, we can see that $S(k)$ is just the image of $S(k-1)$ under some permutation matrix $P(k-1)$, i.e., $S(k) = P(k-1)(S(k-1))$ and so on. As a result, any set $S(k)$ is an image of a finitely many permutations of the initial set $S(0)$; formally $S(k) = P(k-1)\cdots P(1)P(0)(S(0))$. Therefore, we will refer to the set $S(k)$ *as the image of the set* $S(0)$ *under* $\{P(k)\}$ *at time k*. Also, we will refer to the sequence $\{S(k)\}$ as *the trajectory of the set* $S(0)$ *under* $\{P(k)\}$.

In the next lemma, we show that no chain of permutation matrices has absolute infinite property.

Lemma 6.1 *For any permutation chain* $\{P(k)\}$, *there exists a regular sequence* $\{S(k)\}$ *for which*

$$F(\{P(k)\}; \{S(k)\}) = 0.$$

Proof Let $\{P(k)\}$ be an arbitrary permutation chain and let $\{S(k)\}$ be the trajectory of a nonempty set $S(0) \subset [m]$ under the permutation $\{P(k)\}$. Note that $\{S(k)\}$ is regular, and we have

$$P_{S(k+1),S(k)}(k) = \sum_{i \in S(k+1), j \in \bar{S}(k)} P_{ij}(k) + \sum_{i \in \bar{S}(k+1), j \in S(k)} P_{ij}(k) = 0,$$

which is true since $P(k)$ is a permutation matrix and $S(k+1)$ is the image of $S(k)$ under $P(k)$. **Q.E.D.**

Hence, by this lemma none of the permutation sequences $\{P(k)\}$ has absolute infinite flow property.

6.2 Necessity of Absolute Infinite Flow for Ergodicity

As discussed in Theorem 3.1, infinite flow property is necessary for ergodicity of a stochastic chain. In this section, we show that the absolute infinite flow property is actually necessary for ergodicity of a stochastic chain despite the fact that this property is much more restrictive than infinite flow property. We do this by considering a stochastic chain $\{A(k)\}$ and a related chain, say $\{B(k)\}$, such that the flow of $\{A(k)\}$ over a trajectory translates to a flow of $\{B(k)\}$ over an appropriately defined trajectory. The technique that we use for defining the chain $\{B(k)\}$ related to a given chain $\{A(k)\}$ is developed in the following section. Then, we prove the necessity of absolute infinite flow for ergodicity.

6.2.1 Rotational Transformation

Rotational transformation is a process that takes a chain and produces another chain through the use of a permutation sequence $\{P(k)\}$. Specifically, we have the following definition of the rotational transformation with respect to a permutation chain.

Definition 6.3 Given a permutation chain $\{P(k)\}$, the rotational transformation of an arbitrary chain $\{A(k)\}$ with respect to $\{P(k)\}$ is the chain $\{B(k)\}$ given by

$$B(k) = P^T(k+1:0)A(k)P(k:0) \quad \text{for } k \geq 0,$$

where $P(0:0) = I$. We say that $\{B(k)\}$ is the rotational transformation of $\{A(k)\}$ by $\{P(k)\}$.

The rotational transformation has some interesting properties for stochastic chains which we discuss in the following lemma. These properties play a key role in the subsequent development, while they may also be of interest in their own right.

Lemma 6.2 *Let $\{A(k)\}$ be an arbitrary stochastic chain and $\{P(k)\}$ be an arbitrary permutation chain. Let $\{B(k)\}$ be the rotational transformation of $\{A(k)\}$ by $\{P(k)\}$. Then, the following statements are valid:*

(a) *The chain $\{B(k)\}$ is stochastic. Furthermore,*

$$B(k:s) = P^T(k:0)A(k:s)P(s:0) \qquad \text{for any } k > s \geq 0,$$

where $P(0:0) = I$.
(b) *The chain $\{A(k)\}$ is ergodic if and only if the chain $\{B(k)\}$ is ergodic.*
(c) *For any regular sequence $\{S(k)\}$ for $\{A(k)\}$, there exists another regular sequence $\{T(k)\}$ for $\{B(k)\}$, such that $A_{S(k+1)S(k)}(k) = B_{T(k+1)T(k)}(k)$. Also, for any regular sequence $\{T(k)\}$ for $\{B(k)\}$, there exists a regular sequence $\{S(k)\}$ for $\{A(k)\}$, with*

$$A_{S(k+1)S(k)}(k) = B_{T(k+1)T(k)}(k).$$

In particular, $F(\{A(k)\}; \{S(k)\}) = F(\{B(k)\}; \{T(k)\})$ and hence, $\{A(k)\}$ has absolute infinite flow property if and only if $\{B(k)\}$ has absolute infinite flow property.
(d) *For any $S \subset [m]$ and $k \geq 0$, we have $A_{S(k+1),S(k)}(k) = B_S(k)$, where $S(k)$ is the image of S under $\{P(k)\}$ at time k, i.e., $S(k) = P(k:0)(S)$.*

Proof

(a) By the definition of $B(k)$, we have $B(k) = P^T(k+1:0)A(k)P(k:0)$. Thus, $B(k)$ is stochastic as the product of finitely many stochastic matrices is a stochastic matrix.

The proof of relation $B(k:s) = P^T(k:0)A(k:s)P(s:0)$ proceeds by induction on k for $k > s$ and an arbitrary but fixed $s \geq 0$. For $k = s+1$, by the definition of $B(s)$ (see Definition 6.3), we have $B(s) = P^T(s+1:0)A(s)P(s:0)$, while $B(s+1:s) = B(s)$ and $A(s+1:s) = A(s)$. Hence, $B(s+1,s) = P^T(s+1:0)A(s+1:s)P(s:0)$ which shows that $B(k,s) = P^T(k:0)A(k:s)P(s:0)$ for $k = s+1$, thus implying that the claim is true for $k = s+1$.

Now, suppose that the claim is true for some $k > s$, i.e., $B(k,s) = P^T(k:0)A(k:s)P(s:0)$ for some $k > s$. Then, for $k+1$ we have

$$B(k+1:s) = B(k)B(k:s) = B(k)\left(P^T(k:0)A(k:s)P(s:0)\right), \qquad (6.6)$$

where the last equality follows by the induction hypothesis. By the definition of $B(k)$, we have $B(k) = P^T(k+1:0)A(k)P(k:0)$, and by replacing $B(k)$ by $P^T(k+1:0)A(k)P(k:0)$ in Eq. (6.6), we obtain

$$B(k+1:s) = \left(P^T(k+1:0)A(k)P(k:0)\right)\left(P^T(k:0)A(k:s)P(s:0)\right)$$
$$= P^T(k+1:0)A(k)\left(P(k:0)P^T(k:0)\right)A(k:s)P(s:0)$$
$$= P^T(k+1:0)A(k)A(k:s)P(s:0),$$

where the last equality follow from $P^T P = I$ which is valid for any permutation matrix P, and the fact that the product of two permutation matrices is a permutation matrix. Since $A(k)A(k:s) = A(k+1:s)$, it follows that

$$B(k+1:s) = P^T(k+1:0)A(k+1:s)P(s:0),$$

thus showing that the claim is true for $k+1$.

(b) Let the chain $\{A(k)\}$ be ergodic and fix an arbitrary starting time $t_0 \geq 0$. Then, for any $\epsilon > 0$, there exists a sufficiently large time $N_\epsilon \geq t_0$, such that the rows of $A(k:t_0)$ are within ϵ-vicinity of each other; specifically $\|A_i(k:t_0) - A_j(k:t_0)\| \leq \epsilon$ for any $k \geq N_\epsilon$ and all $i, j \in [m]$. We now look at the matrix $B(k:t_0)$ and its rows. By part (a), we have for all $k > t_0$,

$$B(k:t_0) = P^T(k:0)A(k:t_0)P(t_0:0).$$

Furthermore, the ith row of $B(k:t_0)$ can be represented as $e_i^T B(k:t_0)$. Therefore, the norm of the difference between the ith and jth row of $B(k:t_0)$ is given by

$$\|B_i(k:t_0) - B_j(k:t_0)\| = \|(e_i - e_j)^T B(k:t_0)\|$$
$$= \|(e_i - e_j)^T P^T(k:0)A(k:t_0)P(t_0:0)\|.$$

Letting $e_{i(k)} = P(k:0)e_i$ for any $i \in [m]$, we further have

$$\|B_i(k:t_0) - B_j(k:t_0)\| = \|(e_{i(k)} - e_{j(k)})^T A(k:t_0)P(t_0:0)\|$$
$$= \|(A_{i(k)}(k:t_0) - A_{j(k)}(k:t_0))P(t_0:0)\|$$
$$= \|A_{i(k)}(k:t_0) - A_{j(k)}(k:t_0)\|, \qquad (6.7)$$

where the last inequality holds since $P(t_0:0)$ is a permutation matrix and $\|Px\| = \|x\|$ for any permutation P and any $x \in \mathbb{R}^m$. Choosing $k \geq N_\epsilon$ and using

$$\|A_i(k:t_0) - A_j(k:t_0)\| \leq \epsilon,$$

for any $k \geq N_\epsilon$ and all $i, j \in [m]$, we obtain

$$\|B_i(k : t_0) - B_j(k : t_0)\| \leq \epsilon \qquad \text{for any } k \geq N_\epsilon \text{ and all } i, j \in [m].$$

Therefore, it follows that the ergodicity of $\{A(k)\}$ implies the ergodicity of $\{B(k)\}$.

For the reverse implication we note that $A(k : t_0) = P(k : 0)B(k : t_0)P^T(t_0 : 0)$, which follows by part (a) and the fact $PP^T = PP^T = I$ for any permutation P. The rest of the proof follows a line of analysis similar to the preceding case, where we exchange the roles of $B(k : t_0)$ and $A(k : t_0)$.

(c) Let $\{S(k)\}$ be a regular sequence. Let $T(k)$ be the image of $S(k)$ under the permutation $P^T(k : 0)$, i.e., $T(k) = P^T(k : 0)(S(k))$. Note that $|T(k)| = |S(k)|$ for any $k \geq 0$ and since $\{S(k)\}$ is a regular sequence, it follows that $\{T(k)\}$ is a regular sequence. Now, by the definition of rotational transformation we have $A(k) = P(k + 1 : 0)B(k)P^T(k : 0)$, and hence:

$$\sum_{i \in S(k+1), j \in \bar{S}(k)} e_i^T A(k)e_j = \sum_{i \in S(k+1), j \in \bar{S}(k)} e_i^T [P(k + 1 : 0)B(k)P^T(k : 0)]e_j$$

$$= \sum_{i \in T(k+1), j \in \bar{T}(k)} e_i^T B(k)e_j.$$

Similarly, we have $\sum_{i \in \bar{S}(k+1), j \in S(k)} e_i^T A(k)e_j = \sum_{i \in \bar{T}(k+1), j \in T(k)} e_i^T B(k)e_j$. Now, note that

$$A_{S(k+1)S(k)}(k) = \sum_{i \in S(k+1), j \in \bar{S}(k)} e_i^T A(k)e_j + \sum_{i \in \bar{S}(k+1), j \in S(k)} e_i^T A(k)e_j,$$

$$B_{T(k+1)T(k)}(k) = \sum_{i \in T(k+1), j \in \bar{T}(k)} e_i^T B(k)e_j + \sum_{i \in \bar{T}(k+1), j \in T(k)} e_i^T B(k)e_j.$$

Hence, we have $A_{S(k+1)S(k)}(k) = B_{T(k+1)T(k)}(k)$, implying

$$F(\{A(k)\}; \{S(k)\}) = F(\{B(k)\}; \{T(k)\}).$$

For the converse, for any regular sequence $\{T(k)\}$, if we let $S(k) = P(k : 0)(T(k))$ and using the same line of argument, we conclude that $\{S(k)\}$ is a regular sequence and $A_{S(k+1)S(k)}(k) = B_{T(k+1)T(k)}(k)$. Therefore, $F(\{A(k)\}; \{S(k)\}) = F(\{B(k)\}; \{T(k)\})$ and hence, $\{A(k)\}$ has absolute infinite flow property if and only if $\{B(k)\}$ has absolute infinite flow property.

(d) If $\{P(k)\}$ is such that $S(k)$ is the image of S under $\{P(k)\}$ at any time $k \geq 0$, by the previous part, it follows that $A_{S(k+1)S(k)}(k) = B_{T(k+1)T(k)}(k)$, where $T(k) = P^T(k : 0)(S(k))$. But since $S(k)$ is the image of S under $\{P(k)\}$ at time k, it follows that $P^T(k : 0)(S(k)) = S$, which follows from the fact that $P^T(k : 0)P(k : 0) = I$. Similarly, $T(k + 1) = S$ and hence, by the previous part $A_{S(k+1)S(k)}(k) = B_S(k)$.   **Q.E.D.**

As listed in Lemma 6.2, the rotational transformation has some interesting properties: it preserves ergodicity and it preserves absolute infinite flow property. We will use these properties intensively in the development in this section and the rest of this chapter.

6.2.2 Necessity of Absolute Infinite Flow Property

In this section, we establish the necessity of absolute infinite flow property for ergodicity of stochastic chains. The proof of this result relies on Lemma 6.2.

Theorem 6.1 *The absolute infinite flow property is necessary for ergodicity of any stochastic chain.*

Proof Let $\{A(k)\}$ be an ergodic stochastic chain. Let $\{S(k)\}$ be any regular sequence. Then, there is a permutation sequence $\{P(k)\}$ such that $\{S(k)\}$ is the trajectory of the set $S(0) \subset [m]$ under $\{P(k)\}$, i.e., $S(k) = P(k:0)(S(0))$ for all k, where $P(0:0) = I$. Let $\{B(k)\}$ be the rotational transformation of $\{A(k)\}$ by the permutation sequence $\{P(k)\}$. Then, by Lemma 6.2 (a), the chain $\{B(k)\}$ is stochastic. Moreover, by Lemma 6.2 (b), the chain $\{B(k)\}$ is ergodic. Now, by the necessity of infinite flow property (Theorem 3.1), the chain $\{B(k)\}$ should have infinite flow property, i.e.,

$$\sum_{k=0}^{\infty} B_S(k) = \infty \qquad \text{for any } S \subset [m]. \tag{6.8}$$

Therefore, in particular, we must have $\sum_{k=0}^{\infty} B_S(k) = \infty$ for $S = S(0)$. By Lemma 6.2 (d), Eq. (6.8) implies

$$\sum_{k=0}^{\infty} A_{S(k+1),S(k)}(k) = \infty,$$

thus showing that $\{A(k)\}$ has absolute infinite flow property. **Q.E.D.**

The converse statement of Theorem 6.1 is not true generally, namely absolute infinite flow need not be sufficient for ergodicity of a chain. We reinforce this statement later in Sect. 6.3 (Corollary 6.1). Thus, even though absolute infinite flow property requires a lot of structure for a chain $\{A(k)\}$, by requiring that the flow of $\{A(k)\}$ over any regular sequence $\{S(k)\}$ be infinite, this is still not enough to guarantee ergodicity of the chain. However, as we will soon see, it turns out that this property is sufficient for ergodicity of the doubly stochastic chains.

6.3 Decomposable Stochastic Chains

In this section, we consider a class of stochastic chains, termed decomposable, for which verifying absolute infinite flow property can be reduced to showing that the flows over some specific regular sequences are infinite. We explore some properties of this class which will be also used in later sections.

We start with the definition of a decomposable chain.

Definition 6.4 A chain $\{A(k)\}$ is decomposable if $\{A(k)\}$ can be represented as a nontrivial convex combination of a permutation chain $\{P(k)\}$ and a stochastic chain $\{\tilde{A}(k)\}$, i.e., there exists a $\gamma > 0$ such that

$$A(k) = \gamma P(k) + (1 - \gamma)\tilde{A}(k) \qquad \text{for all } k \geq 0. \tag{6.9}$$

We refer to $\{P(k)\}$ as a permutation component of $\{A(k)\}$ and to γ as a mixing coefficient for $\{A(k)\}$.

An example of a decomposable chain is a chain $\{A(k)\}$ that has strong feedback property, i.e., with $A_{ii}(k) \geq \gamma$ for all $k \geq 0$ and some $\gamma > 0$. In this case, $A(k) = \gamma I + (1 - \gamma)\tilde{A}(k)$ where $\tilde{A}(k) = \frac{1}{1-\gamma}(A(k) - \gamma I)$. Note that $A(k) - \gamma I \geq 0$ and $(A(k) - \gamma I)e = (1 - \gamma)e$, which follows from the stochasticity of $A(k)$. Therefore, $\tilde{A}(k)$ is a stochastic matrix for any $k \geq 0$ and the trivial permutation $\{I\}$ is a permutation component of $\{A(k)\}$. Later, we will show that *any* doubly stochastic chain is decomposable.

We have some side remarks about decomposable chains. The first remark is an observation that a permutation component of a decomposable chain $\{A(k)\}$ need not to be unique. An extreme example is the chain $\{A(k)\}$ with $A(k) = \frac{1}{m}ee^T$ for all $k \geq 0$. Since $\frac{1}{m}ee^T = \frac{1}{m!}\sum_{\xi=1}^{m!} P^{(\xi)}$, any sequence of permutation matrices is a permutation component of $\{A(k)\}$. Another remark is about a mixing coefficient γ of a chain $\{A(k)\}$. Note that mixing coefficient is independent of the permutation component. Furthermore, if $\gamma > 0$ is a mixing coefficient for a chain $\{A(k)\}$, then any $\xi \in (0, \gamma]$ is also a mixing coefficient for $\{A(k)\}$, as it can be seen from the decomposition in Eq. (6.9).

An interesting property of any decomposable chain is that if they are rotationally transformed with respect to their permutation component, the resulting chain has trivial permutation component $\{I\}$. This property is established in the following lemma.

Lemma 6.3 *Let $\{A(k)\}$ be a decomposable chain with a permutation* component *$\{P(k)\}$ and a mixing coefficient γ. Let $\{B(k)\}$ be the rotational transformation of $\{A(k)\}$ with respect to $\{P(k)\}$. Then, the chain $\{B(k)\}$ is decomposable with a trivial permutation component $\{I\}$ and a mixing coefficient γ.*

Proof Note that by the definition of a decomposable chain (Definition 6.4), we have

$$A(k) = \gamma P(k) + (1 - \gamma)\tilde{A}(k) \qquad \text{for any } k \geq 0,$$

where $P(k)$ is a permutation matrix and $\tilde{A}(k)$ is a stochastic matrix. Therefore,

$$A(k)P(k : 0) = \gamma P(k)P(k : 0) + (1 - \gamma)\tilde{A}(k)P(k : 0).$$

By noticing that $P(k)P(k : 0) = P(k + 1 : 0)$ and by using left-multiplication with $P^T(k + 1 : 0)$, we obtain

$$P^T(k + 1 : 0)A(k)P(k : 0)$$
$$= \gamma P^T(k + 1 : 0)P(k + 1 : 0) + (1 - \gamma)P^T(k + 1 : 0)\tilde{A}(k)P(k : 0).$$

By the definition of the rotational transformation (Definition 6.3), we have $B(k) = P^T(k + 1 : 0)A(k)P(k : 0)$. Using this and the fact $P^T P = I$ for any permutation matrix P, we further have

$$B(k) = \gamma I + (1 - \gamma)P^T(k + 1 : 0)\tilde{A}(k)P(k : 0).$$

Define $\tilde{B}(k) = P^T(k + 1 : 0)\tilde{A}(k)P(k : 0)$ and note that each $\tilde{B}(k)$ is a stochastic matrix. Hence,

$$B(k) = \gamma I + (1 - \gamma)\tilde{B}(k), \tag{6.10}$$

thus showing that the chain $\{B(k)\}$ is decomposable with the trivial permutation component and a mixing coefficient γ. **Q.E.D.**

In the next lemma, we prove that absolute infinite flow property and infinite flow property are one and the same for decomposable chains with a trivial permutation component.

Lemma 6.4 *For a decomposable chain with a trivial permutation component, infinite flow property and absolute infinite flow property are equivalent.*

Proof By definition, absolute infinite flow property implies infinite flow property for any stochastic chain. For the reverse implication, let $\{A(k)\}$ be decomposable with a permutation component $\{I\}$. Also, assume that $\{A(k)\}$ has infinite flow property. We claim that $\{A(k)\}$ has absolute infinite flow property. To see this, let $\{S(k)\}$ be any regular sequence. If $S(k)$ is constant after some time t_0, i.e., $S(k) = S(t_0)$ for $k \geq t_0$ and some $t_0 \geq 0$, then

$$\sum_{k=t_0}^{\infty} A_{S(k+1),S(k)}(k) = \sum_{k=t_0}^{\infty} A_{S(t_0)}(k) = \infty,$$

where the last equality holds since $\{A(k)\}$ has infinite flow property. Therefore, if $S(k) = S(t_0)$ for $k \geq t_0$, then we must have $\sum_{k=0}^{\infty} A_{S(k+1),S(k)}(k) = \infty$.

If there is no $t_0 \geq 0$ with $S(k) = S(t_0)$ for $k \geq t_0$, then we must have $S(k_r + 1) \neq S(k_r)$ for an increasing time sequence $\{k_r\}$. Now, for an $i \in S(k_r) \setminus S(k_r + 1) \neq \emptyset$, we have $A_{S(k_r+1),S(k_r)}(k_r) \geq A_{ii}(k_r)$ since $i \in \bar{S}(k_r + 1)$. Furthermore, $A_{ii}(k) \geq \gamma$ for all k since $\{A(k)\}$ has the trivial permutation sequence $\{I\}$ as a permutation component with a mixing coefficient γ. Therefore,

$$\sum_{k=0}^{\infty} A_{S(k+1),S(k)}(k) \geq \sum_{r=0}^{\infty} A_{S(k_r+1),S(k_r)}(k_r) \geq \gamma \sum_{r=0}^{\infty} 1 = \infty.$$

All in all, $F(\{A(k)\}, \{S(k)\}) = \infty$ for any regular sequence $\{S(k)\}$ and, hence, the chain $\{A(k)\}$ has absolute infinite flow property.   **Q.E.D.**

Lemma 6.4 shows that absolute infinite flow property may be easier to verify for the chains with a trivial permutation component, by just checking infinite flow property. This result, together with Lemma 6.3 and the properties of rotational transformation established in Lemma 6.2, provide a basis for showing that a similar reduction of absolute infinite flow property is possible for any decomposable chain.

Theorem 6.2 *Let $\{A(k)\}$ be a decomposable chain with a permutation component $\{P(k)\}$. Then, the chain $\{A(k)\}$ has absolute infinite flow property if and only if $F(\{A(k)\}; \{S(k)\}) = \infty$ for any trajectory $\{S(k)\}$ under $\{P(k)\}$, i.e., for all $S(0) \subset [m]$ and its trajectory $\{S(k)\}$ under $\{P(k)\}$.*

Proof Since, by definition, absolute infinite flow property requires $F(\{A(k)\}; \{S(k)\}) = \infty$ for any regular sequence $\{S(k)\}$, it suffice to show that $F(\{A(k)\}; \{S(k)\}) = \infty$ for any trajectory $\{S(k)\}$ under $\{P(k)\}$. To show this, let $\{B(k)\}$ be the rotational transformation of $\{A(k)\}$ with respect to $\{P(k)\}$. Since $\{A(k)\}$ is decomposable, by Lemma 6.3, it follows that $\{B(k)\}$ has the trivial permutation component $\{I\}$. Therefore, by Lemma 6.4 $\{B(k)\}$ has absolute infinite flow property if and only if it has infinite flow property, i.e.,

$$\sum_{k=0}^{\infty} B_S(k) = \infty \qquad \text{for all nonempty } S \subset [m]. \tag{6.11}$$

By Lemma 6.2 (d), we have $B_S(k) = A_{S(k+1),S(k)}(k)$, where $S(k)$ is the image of $S(0) = S$ under the permutation $\{P(k)\}$ at time k. Therefore, Eq. (6.11) holds if and only if

$$\sum_{k=0}^{\infty} A_{S(k+1)S(k)}(k) = \infty,$$

which in view of $F(\{A(k)\}; \{S(k)\}) = \sum_{k=0}^{\infty} A_{S(k+1)S(k)}(k)$ shows that $F(\{A(k)\}; \{S(k)\}) = \infty$.   **Q.E.D.**

In light of Theorem 6.2, verification of absolute infinite flow property for a decomposable chain is considerably simpler than for an arbitrary stochastic chain. For decomposable chains, it suffice to verify $F(\{A(k)\}; \{S(k)\}) = \infty$ only for the trajectory $\{S(k)\}$ of $S(0)$ under a permutation component $\{P(k)\}$ of $\{A(k)\}$ for any $S(0) \subset [m]$.

Another direct consequence of Lemma 6.4 is that absolute infinite flow property is not generally sufficient for ergodicity.

Corollary 6.1 *Absolute infinite flow property is not a sufficient condition for ergodicity.*

Proof Consider the following static chain:

$$A(k) = \begin{bmatrix} 1 & 0 & 0 \\ \frac{1}{3} & \frac{1}{3} & \frac{1}{3} \\ 0 & 0 & 1 \end{bmatrix} \qquad \text{for } k \geq 0.$$

It can be seen that $\{A(k)\}$ has infinite flow property. Furthermore, it can be seen that $\{A(k)\}$ is decomposable and has the trivial permutation sequence $\{I\}$ as a permutation component. Thus, by Lemma 6.4, the chain $\{A(k)\}$ has absolute infinite flow property. However, $\{A(k)\}$ is not ergodic. This can be seen by noticing that the vector $v = (1, \frac{1}{2}, 0)^T$ is a fixed point of the dynamics $x(k+1) = A(k)x(k)$ with $x(0) = v$, i.e., $v = A(k)v$ for any $k \geq 0$. Hence, $\{A(k)\}$ is not ergodic. **Q.E.D.**

Although absolute infinite flow property is a stronger necessary condition for ergodicity than infinite flow property, Corollary 6.1 demonstrates that absolute infinite flow property is not yet strong enough to be equivalent to ergodicity. However, using the results developed so far, we will show that absolute infinite flow property is in fact equivalent to ergodicity for doubly stochastic chains, as discussed in the following section.

6.4 Doubly Stochastic Chains

In this section, we focus on the class of the doubly stochastic chains. We first show that this class is a subclass of the decomposable chains. Using this result and the results developed in the preceding sections, we establish that absolute infinite flow property is equivalent to ergodicity for doubly stochastic chains.

We start our development by proving that a doubly stochastic chain is decomposable. The key ingredient in this development is the Birkhoff-von Neumann theorem (Theorem 2.5).

Consider a sequence $\{A(k)\}$ of doubly stochastic matrices. By applying Birkhoff-von Neumann theorem to each $A(k)$, we have

$$A(k) = \sum_{\xi=1}^{m!} q_\xi(k) P^{(\xi)}, \tag{6.12}$$

where $\sum_{\xi=1}^{m!} q_\xi(k) = 1$ and $q_\xi(k) \geq 0$ for all $\xi \in [m!]$ and $k \geq 0$. Since $\sum_{\xi=1}^{m!} q_\xi(k) = 1$ and $q_\xi(k) \geq 0$, there exists a scalar $\gamma \geq \frac{1}{m!}$ such that for every $k \geq 0$, we can find $\xi(k) \in [m!]$ satisfying $q_{\xi(k)}(k) \geq \gamma$. Therefore, for any time $k \geq 0$, there is a permutation matrix $P(k) = P^{(\xi(k))}$ such that

$$A(k) = \gamma P(k) + \sum_{\xi=1}^{m!} \alpha_\xi(k) P^{(\xi)} = \gamma P(k) + (1 - \gamma)\tilde{A}(k), \tag{6.13}$$

where $\gamma > 0$ is a *time-independent* scalar and $\tilde{A}(k) = \frac{1}{1-\gamma} \sum_{\xi=1}^{m!} \alpha_\xi(k) P^{(\xi)}$.

The decomposition of $A(k)$ in Eq. (6.13) fits the description in the definition of decomposable chains (Definition 6.4). Therefore, we have established the following result.

Lemma 6.5 *Any doubly stochastic chain is a decomposable chain.*

In the light of Lemma 6.5, all the results developed in Sect. 6.3 are applicable to doubly stochastic chains. In particular, Theorem 6.2 is the most relevant, which states that verifying absolute infinite flow property for decomposable chains can be reduced to verifying infinite flow along particular sequences of the index sets. Another result that we use is the special instance of Theorem 6 in [1] as applied to doubly stochastic chains. Any doubly stochastic chain that has the trivial permutation component $\{I\}$ [i.e., Eq. (6.13) holds with $P(k)=I$] fits the framework of Theorem 4.5.

Now, we are ready to deliver our main result of this section, showing that ergodicity and absolute infinite flow are equivalent for doubly stochastic chains. We accomplish this by combining Theorem 6.2 and 4.5.

Theorem 6.3 *A doubly stochastic chain $\{A(k)\}$ is ergodic if and only if it has absolute infinite flow property.*

Proof Let $\{P(k)\}$ be a permutation component for $\{A(k)\}$ and let $\{B(k)\}$ be the rotational transformation of $\{A(k)\}$ with respect to its permutation component. By Lemma 6.3, $\{B(k)\}$ has the trivial permutation component $\{I\}$. Moreover, since $B(k) = P^T(k+1:0)A(k)P(k:0)$, where $P^T(k+1:0)$, $A(k)$ and $P(k:0)$ are doubly stochastic matrices, it follows that $\{B(k)\}$ is a doubly stochastic chain. Therefore, by application of Theorem 6.3 to doubly stochastic chain with strong feedback property, it follows that $\{B(k)\}$ is ergodic if and only if it has infinite flow property. Then, by Lemma 6.2(d), the chain $\{B(k)\}$ has infinite flow property if and only if $\{A(k)\}$ has absolute infinite flow property. **Q.E.D.**

Theorem 6.3 provides an *alternative characterization* of ergodicity for doubly stochastic chains, under only requirement to have absolute infinite flow property. We note that Theorem 6.3 does not impose any other specific conditions on matrices $A(k)$ such as uniformly bounded diagonal entries or uniformly bounded positive entries, which have been typically assumed in the existing literature (see for example [2–8]).

We observe that absolute infinite flow typically requires verifying the existence of infinite flow along every regular sequence of index sets. However, to use Theorem 6.3, we do not have to check infinite flow for every regular sequence. This reduction in checking absolute infinite flow property is due to Theorem 6.2, which shows that in order to assert absolute infinite flow property for doubly stochastic chains, it suffice that the flow over some specific regular sets is infinite. We summarize this observation in the following corollary.

Corollary 6.2 *Let $\{A(k)\}$ be a doubly stochastic chain with a permutation component $\{P(k)\}$. Then, the chain is ergodic if and only if $F(\{A(k)\}; \{S(k)\}) = \infty$ for all trajectories $\{S(k)\}$ of subsets $S(0) \subset [m]$ under $\{P(k)\}$.*

6.4.1 Rate of Convergence

Here, we explore the rate of convergence result for an ergodic doubly stochastic chain $\{A(k)\}$ which is based on the rate of convergence result developed in Chap. 4. The major ingredient in the development is the establishment of another important property of rotational transformation related to the invariance of the Lyapunov function.

Let $\{A(k)\}$ be a doubly stochastic chain and consider a dynamic $\{x(k)\}$ driven by $\{A(k)\}$ starting at some initial condition $(t_0, x(t_0)) \in \mathbb{Z}^+ \times \mathbb{R}^m$. Note that for a doubly stochastic chain, the static sequence $\{\frac{1}{m}e\}$ is an absolute probability sequence. In this case, the associated quadratic comparison function would be a Lyapunov function defined by:

$$V(x) = \frac{1}{m} \sum_{i=1}^{m} (x_i - \bar{x})^2 \qquad \text{for } x \in \mathbb{R}^m, \tag{6.14}$$

where $\bar{x} = \frac{1}{m} e^T x$ is the average of the entries of the vector x.

We now consider the behavior of the Lyapunov function under rotational transformation of the chain $\{A(k)\}$, as given in Definition 6.3. It emerged that the Lyapunov function V is invariant under the rotational transformation, as shown in forthcoming Lemma 6.6. We emphasize that *the invariance of the Lyapunov function V holds for arbitrary stochastic chain*; the doubly stochasticity of the chain is not needed at all.

Lemma 6.6. *Let $\{A(k)\}$ be a stochastic chain and $\{P(k)\}$ be an arbitrary permutation chain. Let $\{B(k)\}$ be the rotational transformation of $\{A(k)\}$ by $\{P(k)\}$. Let $\{x(k)\}$ and $\{y(k)\}$ be the dynamics obtained by $\{A(k)\}$ and $\{B(k)\}$, respectively, with the same initial point $y(0) = x(0)$ where $x(0) \in \mathbb{R}^m$ is arbitrary. Then, for the function $V(\cdot)$ defined in Eq. (6.14) we have*

$$V(x(k)) = V(y(k)) \qquad \text{for all } k \geq 0.$$

Proof Since $\{y(k)\}$ is the dynamics obtained by $\{B(k)\}$, there holds for any $k \geq 0$,

$$y(k) = B(k-1)y(k-1) = \ldots = B(k-1) \cdots B(1)B(0)y(0) = B(k : 0)y(0).$$

By Lemma 6.2 (a), we have $B(k : 0) = P^T(k : 0)A(k : 0)P(0 : 0)$ with $P(0 : 0) = I$, implying

$$y(k) = P^T(k : 0)A(k : 0)y(0) = P^T(k : 0)A(k : 0)x(0) = P^T(k : 0)x(k),$$

$$\tag{6.15}$$

where the second equality follows from $y(0) = x(0)$ and the last equality follows from the fact that $\{x(k)\}$ is the dynamics obtained by $\{A(k)\}$. Now, notice that the function $V(\cdot)$ of Eq. (6.14) is invariant under any permutation, that is $V(Px) = V(x)$ for any permutation matrix P. In view of Eq. (6.15), the vector $y(k)$ is just a permutation of $x(k)$. Hence, $V(y(k)) = V(x(k))$ for all $k \geq 0$. **Q.E.D.**

Consider an ergodic doubly stochastic chain $\{A(k)\}$ with a trivial permutation component $\{I\}$. Let $t_0 = 0$ and for any $\delta \in (0, 1)$ recursively define t_q, as follows:

$$t_{q+1} = \arg \min_{t \geq t_q+1} \min_{S \subset [m]} \sum_{t=t_q}^{t-1} A_S(k) \geq \delta, \tag{6.16}$$

where the second minimum in the above expression is taken over all nonempty subsets $S \subset [m]$. Basically, t_q is the first time $t > t_{q-1}$ when the accumulated flow from $t = t_{q-1}+1$ to $t = t_q$ exceeds δ over every nonempty $S \subset [m]$. We refer to the sequence $\{t_q\}$ as *accumulation times* for the chain $\{A(k)\}$. We observe that, when the chain $\{A(k)\}$ has infinite flow property, then t_q exists for any $q \geq 0$, and any $\delta > 0$.

Now based on the rate of convergence derived in Theorem 5.2, for the sequence of time instances $\{t_q\}$, we have the following rate of convergence result.

Lemma 6.7 *Let $\{A(k)\}$ be an ergodic doubly stochastic chain with a trivial permutation component $\{I\}$ and a mixing coefficient $\gamma > 0$. Also, let $\{x(k)\}$ be the dynamics driven by $\{A(k)\}$ starting at an arbitrary point $x(0)$. Then, for any $q \geq 1$, we have*

$$V(x(t_q)) \leq \left(1 - \frac{\gamma\delta(1-\delta)^2}{m(m-1)^2}\right) V(x(t_{q-1})),$$

where t_q is defined in (6.16).

Proof Note that any deterministic chain can be considered as an independent random chain. Thus, the result follows by letting $p^* = \frac{1}{m}$ and $\epsilon = 1$ in Theorem 5.2. **Q.E.D.**

Using the invariance of the Lyapunov function under rotational transformation and the properties of rotational transformation, we can establish a result analogous to Lemma 6.7 for an arbitrary ergodic chain of doubly stochastic matrices $\{A(k)\}$. In other words, we can extend Lemma 6.7 to the case when the chain $\{A(k)\}$ does not necessarily have trivial permutation component $\{I\}$. To do so, we appropriately adjust the definition of the accumulation times $\{t_q\}$ for this case. In particular, we let $\delta > 0$ be arbitrary but fixed, and let $\{P(k)\}$ be a permutation component of an ergodic chain $\{A(k)\}$. Next, we let $t_0 = 0$ and for $q \geq 1$, we define t_q as follows:

$$t_{q+1} = \arg \min_{t \geq t_q+1} \min_{S(0) \subset [m]} \sum_{t=t_q^\delta}^{t-1} A_{S(k+1)S(k)}(k) \geq \delta, \tag{6.17}$$

where $\{S(k)\}$ is the trajectory of the set $S(0)$ under $\{P(k)\}$.

We have the following convergence result.

Theorem 6.4. *Let $\{A(k)\}$ be an ergodic doubly stochastic chain with a permutation component $\{P(k)\}$ and a mixing coefficient $\gamma > 0$. Also, $\{x(k)\}$ be the dynamics driven by $\{A(k)\}$ starting at an arbitrary point $x(0)$. Then, for any $q \geq 1$, we have*

$$V(x(t_q)) \leq \left(1 - \frac{\gamma\delta(1-\delta)^2}{m(m-1)^2}\right) V(x(t_{q-1})),$$

where t_q is defined in (6.17).

Proof Let $\{B(k)\}$ be the rotational transformation of the chain $\{A(k)\}$ with respect to $\{P(k)\}$. Also, let $\{y(k)\}$ be the dynamics driven by chain $\{B(k)\}$ with the initial point $y(0) = x(0)$. By Lemma 6.3, $\{B(k)\}$ has the trivial permutation component $\{I\}$. Thus, by Lemma 6.7, we have for all $q \geq 1$,

$$V(y(t_q)) \leq \left(1 - \frac{\gamma\delta(1-\delta)^2}{m(m-1)^2}\right) V(y(t_{q-1})).$$

Now, by Lemma 6.2 (d), we have $A_{S(k+1)S(k)}(k) = B_S(k)$. Therefore, the accumulation times for the chain $\{A(k)\}$ are the same as the accumulation times for the chain $\{B(k)\}$. Furthermore, according Lemma 6.6, we have $V(y(k)) = V(x(k))$ for any $k \geq 0$ and, hence, for all $q \geq 1$,

$$V(x(t_q)) \leq \left(1 - \frac{\gamma\delta(1-\delta)^2}{m(m-1)^2}\right) V(x(t_{q-1})).$$

Q.E.D.

6.4.2 Doubly Stochastic Chains Without Absolute Infinite Flow Property

So far we have been concerned with doubly stochastic chains with absolute infinite flow property. In this section, we turn our attention to the case when absolute flow property is absent. In particular, we are interested in characterizing the limiting behavior of backward product of a doubly stochastic chain that does not have absolute infinite flow.

Let us extend the notion of the infinite flow graph as introduced in Definition 3.3.

Definition 6.5. Let us define the infinite flow graph of $\{A(k)\}$ with respect to a permutation sequence $\{P(k)\}$ to be an undirected graph $G_P^\infty = ([m], \mathcal{E}_P^\infty)$ with

$$\mathcal{E}_P^\infty = \left\{\{i, j\} \,\Big|\, \sum_{k=0}^{\infty} \left(A_{i(k+1),j(k)}(k) + A_{j(k+1)i(k)}(k)\right) = \infty\right\},$$

where $\{i(k)\}$ and $\{j(k)\}$ are the trajectories of the sets $S(0) = \{i\}$ and $S(0) = \{j\}$, respectively, under the permutation component $\{P(k)\}$; formally, $e_{i(k)} = P(k:0)e_i$ and $e_{j(k)} = P(k:0)e_j$ for all k with $P(0:0) = I$. We refer to the graph $G_P^\infty = ([m], \mathcal{E}_P^\infty)$ as the infinite flow graph of the chain $\{A(k)\}$ with respect to a permutation sequence $\{P(k)\}$.

Notice that if we let the permutation sequence $\{P(k)\}$ be the trivial permutation sequence $\{I\}$, then G_P^∞ is nothing but the infinite flow graph G^∞ as given in Definition 3.3.

As discussed in Lemma 3.3, we can state Theorem 3.1 in terms of the infinite flow graph. We can use the same line of argument to restate Theorem 6.1 in terms of the associated infinite flow graphs G_P^∞.

Lemma 6.8. *Let $\{A(k)\}$ be an ergodic chain. Then the infinite flow graph of $\{A(k)\}$ with respect to any permutation sequence $\{P(k)\}$ is connected.*

Proof Let G_P^∞ be an infinite flow graph with respect to a permutation chain $\{P(k)\}$ that is not connected. Let $S \subset [m]$ be a connected component of G_P^∞. Then, we have

$$\sum_{k=0}^{\infty} \sum_{i \in S, j \in \bar{S}} \left(A_{i(k+1)j(k)}(k) + A_{i(k)j(k+1)}(k) \right) < \infty.$$

But $\sum_{i \in S, j \in \bar{S}} A_{i(k)j(k)}(k) = A_{S(k+1)S(k)}(k)$ which implies that $F(\{A(k)\}; \{S(k)\}) < \infty$ and hence, by Theorem 6.1 it follows that $\{A(k)\}$ is not ergodic. **Q.E.D.**

Since a doubly stochastic chain is decomposable, Theorem 6.2 is applicable, so by this theorem when the chain $\{A(k)\}$ does not have absolute infinite flow property, then

$$F(\{A(k)\}; \{S(k)\}) < \infty$$

for some $S(0) \subset [m]$ and its trajectory under a permutation component $\{P(k)\}$ of $\{A(k)\}$. This permutation component will be important so we denote it by $\mathcal{P}$. By Theorem 6.2, we have that $G_{\mathcal{P}}^\infty$ is connected if and only if the chain has absolute infinite flow property. Since, a doubly stochastic chain $\{A(k)\}$ with the trivial permutation component is a chain in $\mathcal{P}^*$ with feedback property, Theorem 4.5 shows the connectivity of $G_{\mathcal{P}}^\infty$ is closely related to the limiting matrices of the product $A(k : t_0)$, as $k \to \infty$.

Using Lemma 6.2, 6.3 and Theorem 6.3, we can show that the backward product of any doubly stochastic chain *essentially* converges.

Theorem 6.5 *Let $\{A(k)\}$ be a doubly stochastic chain with a permutation component $\{P(k)\}$. Then, for any starting time $t_0 \geq 0$, the product $A(k : t_0)$ converges up to a permutation of its rows; i.e., there exists a permutation sequence $\{Q(k)\}$ such that $\lim_{k \to \infty} Q(k)A(k : t_0)$ exists for any $t_0 \geq 0$. Moreover, for the trajectories $\{i(k)\}$ and $\{j(k)\}$ of $S(0) = \{i\}$ and $S(0) = \{j\}$, respectively, under the permutation component $\{P(k)\}$, we have*

$$\lim_{k \to \infty} \| A_{i(k)}(k : t_0) - A_{j(k)}(k : t_0) \| = 0 \qquad \text{for any starting time } t_0,$$

if and only if i and j belong to the same connected component of $G_{\mathcal{P}}^\infty$.

Proof Let $\{B(k)\}$ be the rotational transformation of $\{A(k)\}$ by the permutation component $\{P(k)\}$. As proven in Lemma 6.3, the chain $\{B(k)\}$ has a trivial permutation component. Hence, by Theorem 6.3, the limit $B^\infty(t_0) = \lim_{k\to\infty} B(k : t_0)$ exists for any $t_0 \geq 0$. On the other hand, by Lemma 6.2 (a), we have

$$B(k : t_0) = P^T(k : 0)A(k : t_0)P(t_0 : 0) \qquad \text{for all } k > t_0.$$

Multiplying by $P^T(t_0 : 0)$ from the right, and using $PP^T = I$ which is valid for any permutation matrix P, we obtain

$$B(k : t_0)P(t_0 : 0)^T = P^T(k : 0)A(k : t_0) \qquad \text{for all } k > t_0.$$

Therefore, $\lim_{k\to\infty} B(k : t_0)P(t_0 : 0)^T$ always exists for any starting time t_0 since $B(k : t_0)P^T(t_0 : 0)$ is obtained by a fixed permutation of the columns of $B(k : t_0)$. Therefore, if we let $Q(k) = P^T(k : 0)$, then $\lim_{k\to\infty} Q(k)A(k : t_0)$ exists for any t_0 which proves the first part of the theorem.

For the second part, by Theorem 6.3, we have $\lim_{k\to\infty} \|B_i(k : t_0) - B_j(k : t_0)\| = 0$ for any $t_0 \geq 0$ if and only if i and j belong to the same connected component of the infinite flow graph of $\{B(k)\}$. By the definition of the rotational transformation, we have $B(k : t_0) = P^T(k : 0)A(k : t_0)P(t_0 : 0)$. Therefore, for the ith and jth row of $B(k : t_0)$, we have according to Eq. (6.7):

$$\|B_i(k : t_0) - B_j(k : t_0)\| = \|A_{i(k)}(k : t_0) - A_{j(k)}(k : t_0)\|,$$

where $e_{i(k)} = P(k : 0)e_i$ and $e_{j(k)} = P(k : 0)e_j$ for all k. Therefore, $\lim_{k\to\infty} \|B_i(k : t_0) - B_j(k : t_0)\| = 0$ if and only if $\lim_{k\to\infty} \|A_{i(k)}(k : t_0) - A_{j(k)}(k : t_0)\| = 0$. Thus, $\lim_{k\to\infty} \|A_{i(k)}(k : t_0) - A_{j(k)}(k : t_0)\| = 0$ for any $t_0 \geq 0$ if and only if i and j belong to the same connected component of the infinite flow graph of $\{B(k)\}$. The last step is to show that the infinite flow graph of $\{B(k)\}$ (with respect to the trivial permutation chain) and the infinite flow graph of the chain $\{A(k)\}$ with respect to $\mathcal{P}$ are the same. This however, follows from the following relations:

$$\sum_{k=0}^{\infty} B_{ij}(k) = \sum_{k=0}^{\infty} e_i^T P^T(k+1 : 0)A(k)P(k : 0)e_j = \sum_{k=0}^{\infty} e_{i(k)}A(k)e_{j(k)}.$$

Q.E.D.

By Theorem 6.5, for any doubly stochastic chain $\{A(k)\}$ and any fixed t_0, the sequence consisting of the rows of $A(k : t_0)$ converges to a *multiset of m points* in the probability simplex of $\mathbb{R}^m$, as k approaches to infinity. In general, this is not true for an arbitrary stochastic chain. For example, consider the stochastic chain

$$A(2k) = \begin{bmatrix} 1 & 0 & 0 \\ 1 & 0 & 0 \\ 0 & 0 & 1 \end{bmatrix}, \quad A(2k+1) = \begin{bmatrix} 1 & 0 & 0 \\ 0 & 0 & 1 \\ 0 & 0 & 1 \end{bmatrix} \qquad \text{for all } k \geq 0.$$

For this chain, we have $A(2k : 0) = A(2k)$ and $A(2k + 1 : 0) = A(2k + 1)$. Hence, depending on the parity of k, the set consisting of the rows of $A(k : 0)$ alters between $\{(1, 0, 0), (1, 0, 0), (0, 0, 1)\}$ and $\{(1, 0, 0), (0, 0, 1), (0, 0, 1)\}$ and, hence, never converges to a multiset with three elements in $\mathbb{R}^3$.

References

1. B. Touri, A. Nedić, On ergodicity, infinite flow and consensus in random models. IEEE Trans. Autom. Control **56**(7), 1593–1605 (2011)
2. J. Tsitsiklis, Problems in decentralized decision making and computation, Ph.D. dissertation, Department of Electrical Engineering and Computer Science, MIT, 1984
3. A. Jadbabaie, J. Lin, S. Morse, Coordination of groups of mobile autonomous agents using nearest neighbor rules. IEEE Trans. Autom. Control **48**(6), 988–1001 (2003)
4. M. Cao, A.S. Morse, B.D.O. Anderson, Reaching a consensus in a dynamically changing environment: A graphical approach. SIAM J. Control Optim. **47**, 575–600 (2008)
5. R. Carli, F. Fagnani, A. Speranzon, S. Zampieri, Communication constraints in the average consensus problem. Automatica **44**(3), 671–684 (2008)
6. R. Hegselmann, U. Krause, Opinion dynamics and bounded confidence models, analysis, and simulation. J. Artif. Societies Social Simul. **5** (2002)
7. S. Boyd, A. Ghosh, B. Prabhakar, D. Shah, Randomized gossip algorithms. IEEE Trans. Inf. Theory **52**(6), 2508–2530 (2006)
8. A. Nedić, A. Olshevsky, A. Ozdaglar, J. Tsitsiklis, On distributed averaging algorithms and quantization effects, in *Proceedings of the 47th IEEE Conference on Decision and Control* 2008

Chapter 7
Averaging Dynamics in General State Spaces

Motivated by the theory of Markov chains over general state spaces [1], in this chapter we provide a framework for the study of averaging dynamics over general state spaces. We will show that some of the developed results in the previous chapters remain to be true for arbitrary state spaces. As in Chap. 6, our discussion on general state spaces will be restricted to deterministic chains.

The structure of this chapter is as follows: In Sect. 7.1, we introduce and discuss averaging dynamics over general state spaces. Then, in Sect. 7.2 we discuss several modes of ergodicity. We show that unlike averaging dynamics on $\mathbb{R}^m$, there are several notions of ergodicity for averaging dynamics in a general state space that are not necessarily equivalent. In Sect. 7.3, we discuss the generalization of infinite flow property over a general state space and prove that it is necessary for the weakest form of ergodicity in an arbitrary state space. Finally, in Sect. 7.4 we prove a generalization of the fundamental relation (4.8) for the averaging dynamics in general state spaces.

7.1 Framework

As discussed in Chap. 2, for averaging dynamics in $\mathbb{R}^m$ we are interested in limiting behavior of the dynamics

$$x(k + 1) = A(k)x(k), \text{ for } k \geq t_0, \tag{7.1}$$

where, $(t_0, x(t_0)) \in \mathbb{Z}^+ \times \mathbb{R}^m$ is an initial condition for the dynamics. To distinguish between averaging dynamics in $\mathbb{R}^m$ and general state spaces, we refer to the dynamics in Eq. 7.1 as the classic averaging dynamics. By our earlier discussions in Chap. 2, study of limiting behavior of the dynamics (7.1) is an alternative way of studying the convergence properties of the product $A(k : t_0)$. This viewpoint leads to our operator theoretic viewpoint to averaging dynamics over general state spaces. Also, as it is shown in Theorem 2.2, it suffice to verify the convergence of $\{x(k)\}$ only for $x(t_0) = e_\ell$ where $\ell \in [m]$ which shows that the ℓth column of $A(k : t_0)$ converges.

B. Touri, *Product of Random Stochastic Matrices and Distributed Averaging*,
Springer Theses, DOI: 10.1007/978-3-642-28003-0_7,
© Springer-Verlag Berlin Heidelberg 2012

In our development, we will visit a counterpart of such a result for averaging dynamics in general state spaces.

Before formulating averaging dynamics over general state spaces, let us discuss the notation used in this chapter which is slightly different from the notation used in the previous chapters. Here, instead of sequences of stochastic matrices, we are dealing with sequences of stochastic kernels in a measure space. Since in this case the state variables are more involved in our developments, we use the subscripts for indexing time variables. Thus instead of using the notation $\{\mathcal{K}(k)\}$ for a sequence of stochastic kernels, we use $\{\mathcal{K}_k\}$ to denote a sequence of kernels. Also, we use $\mathcal{K}_k(\xi, \eta)$ to denote the value of $\mathcal{K}_k$ at the point (ξ, η). However, for averaging dynamics over $\mathbb{R}^m$, we still use the same notation as in the previous chapters. This also helps us to distinguish between averaging dynamics over $\mathbb{R}^m$ and averaging dynamics over general state spaces.

To formulate averaging dynamics over a general state space, let X be a set with a σ-algebra $\mathcal{M}$ of subsets of X. Throughout this chapter, the measurable space $(X, \mathcal{M})$ will serve as our general state space.

Definition 7.1 [1] We say a function $\mathcal{K} : X \times \mathcal{M} \to \mathbb{R}^+$ is a stochastic kernel if

(a) for any $S \in \mathcal{M}$, the function $f_S : X \to \mathbb{R}^+$ defined by $f_S(\xi) = \mathcal{K}(\xi, S)$ is a measurable function,
(b) for any $\xi \in X$, the set function $\mathcal{K}(\xi, \cdot)$ is a measure on X,
(c) for any $\xi \in X$, we have $\mathcal{K}(\xi, X) = 1$, i.e. the measure $\mathcal{K}(\xi, \cdot)$ is a probability measure for any $\xi \in X$.

Furthermore, if we can write

$$\mathcal{K}(\xi, S) = \int_S \tilde{\mathcal{K}}(\xi, \eta) d\mu(\eta),$$

for some measurable function $\tilde{\mathcal{K}} : X \times X \to \mathbb{R}^+$ and a measure μ on $(X, \mathcal{M})$, then $\mathcal{K}$ is referred as a stochastic integral kernel with density $\tilde{\mathcal{K}}$ and basis μ.

Let us define L_∞ to be the space of all measurable functions x from $(X, \mathcal{M})$ to $\mathbb{R}$ such that $\sup_{\xi \in X} |x(\xi)| < \infty$, and let $\|x\|_\infty = \sup_{\xi \in X} |x(\xi)|$.

For a chain $\{\mathcal{K}_k\}$ of stochastic kernels, a given starting time $t_0 \geq 0$, and a starting point $x_{t_0} \in L_\infty$, let us define the averaging dynamics as follows:

$$x_{k+1}(\xi) = \int_X \mathcal{K}_k(\xi, d\eta) x_k(\eta) \qquad \text{for any } \xi \in X \text{ and } k \geq t, \tag{7.2}$$

where $\int_X \mathcal{K}_k(\xi, d\eta) x_k(\eta)$ is the integral of x_k with respect to the measure $\mathcal{K}(\xi, \cdot)$. Let us represent Eq. 7.2 concisely by $x_{k+1} = \mathcal{K}_k x_k$.

Note that, when the kernel $\mathcal{K}_k$ in (7.2) is an integral kernel with density $\tilde{\mathcal{K}}_k$ and a basis μ, then we can write (7.2) as

$$x_{k+1}(\xi) = \int_X \tilde{\mathcal{K}}_k(\xi, \eta) x_k(\eta) d\mu(\eta). \tag{7.3}$$

We now show that the dynamics (7.2) is well-defined, in the sense that $\{x(k)\} \subset L_\infty$ whenever the dynamics is started at a point in L_∞.

Lemma 7.1 *Let $\{x(k)\}$ be the dynamics generated by the dynamics (7.2) started at time $t_0 \geq 0$ and the point $x_{t_0} \in L_\infty$. Then, $\|x_{k+1}\|_\infty \leq \|x_k\|_\infty$ for any $k \geq 0$. Thus, $\{x(k)\}$ is a sequence in L_∞.*

Proof Note that $x_{t_0} \in L_\infty$ and by induction for any $\xi \in X$, we have

$$|x_{k+1}(\xi)| = |\int_X \mathcal{K}_k(\xi, d\eta)x_k| \leq \int_X |\mathcal{K}_k(\xi, d\eta)x_k| \leq \|x_k\|_\infty,$$

which holds since $\mathcal{K}_k(\xi, \cdot)$ is a probability measure. Therefore, it follows that $|x_{k+1}(\xi)| \leq \|x_{t_0}\|_\infty$ for all $\xi \in X$ and hence, $x_k \in L_\infty$ for all $k \geq t_0$.   **Q.E.D**

***Example* 7.1** Let $X = [m] = \{1, \ldots, m\}$, $\mathcal{M} = \mathscr{P}([m])$ (the set of all subsets of $[m]$) and let μ be the counting measure on $(X, \mathcal{M})$, i.e. $\mu(S) = |S|$ for any $S \subseteq [m]$. For a chain of stochastic matrices $\{A(k)\}$, if we define $\tilde{\mathcal{K}}_k(i, j) = A_{ij}(k)$ for all $i, j \in [m]$ and $k \geq 0$, then $\{\tilde{\mathcal{K}}_k\}$ is a chain of density functions with basis μ. For such a chain any vector $x_{t_0} \in \mathbb{R}^m$ is in L_∞ and hence, the classic averaging dynamics (7.1) is a special case of the averaging dynamics (7.3) in general state spaces.   **Q.E.D**

7.2 Modes of Ergodicity

In this section, we define several modes of ergodicity for averaging dynamics in (7.2). As in the case of stochastic matrices, a stochastic kernel $\mathcal{K}$ can be viewed as an operator $T : L_\infty \to L_\infty$ defined by $T(x) = \mathcal{K}x$. Note that by Lemma 7.1, we have $\|\mathcal{K}x\|_\infty \leq \|x\|_\infty$. Also, since $\mathcal{K}$ is a stochastic kernel, we have $\mathcal{K}\mathbf{1}_X = \mathbf{1}_X$. On the other hand, by the linearity of integral, we have $\mathcal{K}(x + y) = \mathcal{K}x + \mathcal{K}y$ for any $x, y \in L_\infty$. Thus, $\mathcal{K}$ can be viewed as an element in $B(L_\infty, L_\infty)$ where $B(L_\infty, L_\infty)$ is the set of bounded linear operators from L_∞ to L_∞. Furthermore, $\|\mathcal{K}\|_\infty = 1$ where $\|\mathcal{K}\|_\infty$ is the induced operator norm of $\mathcal{K}$.

A viewpoint to averaging dynamics (7.2) is to view $\{x(k)\}$ as the image of a point x_{t_0} under a sequence of operators $\mathcal{K}_{k:t_0} = \mathcal{K}_{k-1}\mathcal{K}_{k-2} \cdots \mathcal{K}_{t_0}$ as k varies from $t_0 + 1$ to ∞ where PQ should be understood as:

$$[PQ](\xi, S) = \int P(\xi, d\psi)Q(\psi, S),$$

for any $\xi \in X$ and $S \in \mathcal{M}$.

For a probability measure π on $(X, \mathcal{M})$, let us denote the stochastic kernel $\mathcal{K}(\xi, S) = \pi(S)$ by $\mathcal{K} = \mathbf{1}_X \pi^T$.

Definition 7.2 Let $\{\mathcal{K}_k\}$ be a chain of stochastic kernels on $(X, \mathcal{M})$. Then, we say $\{\mathcal{K}_k\}$ is

- Uniformly Ergodic: if $\lim_{k\to\infty} \|\mathcal{K}_{k:t_0} - \mathbf{1}_X \pi_{t_0}^T\|_\infty = 0$ for some probability measure π_{t_0} on $(X, \mathcal{M})$, and any $t_0 \geq 0$, where the equality should be understood as the induced operator norm.
- Strongly Ergodic: if for any set $\xi \in X$, we have $\lim_{k\to\infty} x_k(\xi) = c(x_{t_0}, t_0)$ for some $c(x_{t_0}, t_0) \in \mathbb{R}$ and any initial condition $(t_0, x_{t_0}) \in \mathbb{Z}^+ \times L_\infty$.
- Weakly Ergodic: if for any $\xi, \eta \in X$, we have

$$\lim_{k\to\infty} (x_k(\xi) - x_k(\eta)) = 0,$$

for any initial condition $(t_0, x_{t_0}) \in \mathbb{Z}^+ \times L_\infty$.

Based on Definition 7.2 one can define several modes of consensus types.

Definition 7.3 We say that $\{\mathcal{K}_k\}$ admits Consensus

- Uniformly: if $\lim_{k\to\infty} \|\mathcal{K}_{k:0} - \mathbf{1}_X \pi^T\|_\infty = 0$, for some probability measure π on $(X, \mathcal{M})$.
- Strongly: if for any $\xi \in X$, we have $\lim_{k\to\infty} x_k(\xi) = c(x_0)$ for some $c(x_0) \in \mathbb{R}$ and any starting point $x_0 \in L_\infty$.
- Weakly: if for any $\xi, \eta \in X$, we have $\lim_{k\to\infty} (x_k(\xi) - x_k(\eta)) = 0$ for any starting point $x_0 \in L_\infty$.

For classic averaging dynamics, all these notions of ergodicity and consensus as given in Definition 7.2 and Definition 7.3 are equivalent. However, in general state spaces they lead to different properties.

From the Definition 7.2, it follows that uniform ergodicity implies strong ergodicity which itself implies weak ergodicity. Similar relations hold among the modes of consensus. However, the reverse implications do not necessary hold. The following example shows that in general, strong ergodicity (consensus) does not imply uniform ergodicity (consensus).

Example 7.2 Consider the set of non-negative integers $\mathbb{Z}^+$, and let $\mathcal{M} = \mathscr{P}(\mathbb{Z}^+)$, and μ be the counting measure on $\mathbb{Z}^+$. Let $\{\mathcal{K}_k\}$ be the chain of stochastic kernels with density functions given by

$$\tilde{\mathcal{K}}_k(i, j) = \begin{cases} \delta_{0j} & \text{if } i \leq k, \\ \delta_{ij} & \text{if } i > k, \end{cases}$$

for $k \geq 1$ and for $k = 0$, let $\tilde{\mathcal{K}}_0(i, j) = \delta_{ij}$, where $\delta_{ij} = 1$ of $i = j$ and otherwise, $\delta_{ij} = 0$. In this case, $\tilde{\mathcal{K}}_k$ can be viewed as a $|\mathbb{Z}^+| \times |\mathbb{Z}^+|$ stochastic matrix with the form:

$$
\tilde{\mathcal{K}}_k =
\begin{bmatrix}
1 & \cdots & 0 & 0 & 0 & 0 & \cdots \\
\vdots & \ddots & \vdots & \vdots & \vdots & \vdots & \cdots \\
1 & \cdots & 0 & 0 & 0 & 0 & \cdots \\
1 & \cdots & 0 & 0 & 0 & 0 & \cdots \\
0 & \cdots & 0 & 0 & 1 & 0 & \cdots \\
0 & \cdots & 0 & 0 & 0 & 1 & \cdots \\
\vdots & \vdots & & \vdots & \vdots & \vdots & \vdots & \ddots
\end{bmatrix}.
\tag{7.4}
$$

Now, let $\{x(k)\}$ be a dynamics generated by $\{\mathcal{K}_k\}$ started at an arbitrary initial condition $(t_0, x_{t_0}) \in \mathbb{Z}^+ \times L_\infty$. Then for any $\xi \in \mathbb{Z}^+$, if we let $k > \max(t_0, \xi)$, we have $x_k(\xi) = x_{t_0}(0)$. Thus, $\lim_{k \to \infty} x_k(\xi) = x_{t_0}(0)$. Therefore, $\{\mathcal{K}_k\}$ is strongly ergodic and admits consensus strongly.

However, by the form of Eq. 7.4, the only candidate for $\lim_{k \to \infty} \mathcal{K}_{k:t_0}$ is the integral kernel $\mathbf{1}_X \pi^T$ where π is the probability measure concentrated at $\{0\}$, i.e. $\pi(S) = 1$ if $0 \in S$ and $\pi(S) = 0$, otherwise. However, if we define x_{t_0} by $x_{t_0}(i) = \delta_{0i}$, then we have $\|(\mathcal{K}_k - \mathbf{1}_X \pi^T)x_{t_0}\|_\infty = 1$ and hence, $\{\mathcal{K}_k\}$ is not uniformly ergodic and does not admit consensus uniformly.

The following example shows that weak ergodicity (consensus) does not imply strong ergodicity (consensus).

Example **7.3** Consider the measure space $(\mathbb{Z}^+, \mathcal{M})$, defined in the previous example. Consider the chain of stochastic integral kernels $\{\mathcal{K}_k\}$ with density kernels,

$$
\tilde{\mathcal{K}}_k(i, j) =
\begin{cases}
\delta_{jk} & \text{if } i \leq k, \\
\delta_{ij} & \text{if } i > k
\end{cases}
\qquad \text{for } k \geq 0.
$$

The density kernel $\tilde{\mathcal{K}}_k$ has the following form

$$
\tilde{\mathcal{K}}_k =
\begin{bmatrix}
0 & \cdots & 0 & 1 & 0 & 0 & \cdots \\
\vdots & \ddots & \vdots & \vdots & \vdots & \vdots & \cdots \\
0 & \cdots & 0 & 1 & 0 & 0 & \cdots \\
0 & \cdots & 0 & 1 & 0 & 0 & \cdots \\
0 & \cdots & 0 & 0 & 1 & 0 & \cdots \\
0 & \cdots & 0 & 0 & 0 & 1 & \cdots \\
\vdots & \vdots & & \vdots & \vdots & \vdots & \vdots & \ddots
\end{bmatrix}.
$$

For any two $\xi, \eta \in \mathbb{Z}^+$, and any starting point $x_{t_0} \in L_\infty$, we have $x_k(\xi) = x_k(\eta) = x_k(k)$, for $k \geq K = \max(\xi, \eta)$ and hence, $\lim_{k \to \infty}(x_k(\xi) - x_k(\eta)) = 0$. Nevertheless, if we let $\{\alpha_k\}$ be a sequence of scalars in $[0, 1]$ that is not convergent and we start the dynamics at time $t = 0$ and starting point $x_0 = (\alpha_0, \alpha_1, \alpha_2, \ldots)$ (which belongs to L_∞), then we have $x_k = (\alpha_k, \alpha_k, \ldots, \alpha_k, \alpha_{k+1}, \alpha_{k+2} \ldots)$. Therefore, for any $\xi \in \mathbb{Z}^+$, we have $x_k(\xi) = \alpha_k$ for $k > \xi$. Since $\{\alpha_k\}$ is not convergent, we conclude that $\lim_{k \to \infty} x_k(\xi)$ does not exist and hence, the chain $\{\mathcal{K}_k\}$ is not strongly ergodic and does not admit consensus uniformly.

The definition of weak ergodicity and consensus are inspired by the dynamic system viewpoint to ergodicity and consensus presented in Theorem 2.2 and Theorem 2.3. These properties would be of more interest from the consensus-seeking algorithms and protocols.

7.3 Infinite Flow Property in General State Spaces

In Chap. 3, we defined the infinite flow property and we showed that this property is necessary for ergodicity of classic averaging dynamics. In this section, we show a similar result for averaging dynamics in a general state space.

For a stochastic chain $\{A(k)\}$, the infinite flow property requires that $\sum_{k=0}^{\infty} A_S(k) = \infty$ for any non-trivial $S \subset [m]$ where

$$A_S(k) = A_{S\bar{S}}(k) + A_{\bar{S}S}(k) = \sum_{i \in S, j \in \bar{S}} A_{ij}(k) + \sum_{i \in \bar{S}, j \in S} A_{ij}(k).$$

In a general state space $(X, \mathcal{M})$ that is not equipped with a measure, $\mathcal{K}(S, \bar{S})$ is not well-defined for a stochastic kernel $\mathcal{K}$. However, if we have an integral kernel $\mathcal{K}$ with density $\tilde{\mathcal{K}}$ and a basis μ, we can define

$$\mathcal{K}(S, \bar{S}) = \int_S \mathcal{K}(\xi, \bar{S}) d\mu(\xi) = \int_S \int_{\bar{S}} \tilde{\mathcal{K}}(\xi, \eta) d\mu(\eta) d\mu(\xi).$$

Thus, for a chain of stochastic integral kernels $\{\mathcal{K}_k\}$ with density kernels $\{\tilde{\mathcal{K}}_k\}$ and basis μ, it is tempting to define the flow over a non-trivial set $S \in \mathcal{M}$ (i.e. $S \neq \emptyset$ and $S \neq X$ a.e.) to be

$$\bar{\mathcal{K}}^f(S) = \mathcal{K}(S, \bar{S}) + \mathcal{K}(\bar{S}, S)$$

$$= \int_S \int_{\bar{S}} \tilde{\mathcal{K}}(\xi, \eta) d\mu(\xi) d\mu(\eta) + \int_{\bar{S}} \int_S \tilde{\mathcal{K}}(\xi, \eta) d\mu(\xi) d\mu(\eta), \qquad (7.5)$$

and conclude the necessity of the infinite flow for (at least) uniform ergodicity of any chain of stochastic integral kernels. However, with the definition of flow in Eq. 7.5, such a result is not true as it can be deduced from the following example.

Example **7.4** Let $X = [0, 1]$ with $\mathcal{M}$ being Borel sets of $[0, 1]$ and μ being the Borel-measure on $[0, 1]$. Let the densities $\tilde{\mathcal{K}}_k$ be defines by

$$\tilde{\mathcal{K}}_k(\xi, \eta) = \begin{cases} 2 \cdot \mathbf{1}_{(\frac{1}{2}, 1]}(\eta) & \xi \in [0, 2^{-k}] \\ 2^k \cdot \mathbf{1}_{[0, 2^{-k}]}(\eta) & \xi \in (2^{-k}, \frac{1}{2}] \\ 2 \cdot \mathbf{1}_{(\frac{1}{2}, 1]}(\eta) & \xi \in (\frac{1}{2}, 1] \end{cases} \qquad \text{for } k \geq 0. \qquad (7.6)$$

First, let us show that $\{\mathcal{K}_k\}$ is uniformly ergodic. To do so, we show that $\tilde{\mathcal{K}}_{k+1}\tilde{\mathcal{K}}_k(\xi, \eta) = 2 \cdot \mathbf{1}_{(\frac{1}{2}, 1]}(\eta)$ for any $k \geq 0$ and $\xi \in [0, 1]$. To prove this, consider the following cases:

(i) $\xi \in [0, 2^{-(k+1)}]$: In this case, by Eq. 7.6, we have $\tilde{\mathcal{K}}_{k+1}(\xi, \psi) = 2 \cdot \mathbf{1}_{(\frac{1}{2}, 1]}(\psi)$ and hence,

$$
\tilde{\mathcal{K}}_{k+1}\tilde{\mathcal{K}}_k(\xi, \eta) = \int_{[0,1]} \tilde{\mathcal{K}}_{k+1}(\xi, \psi)\tilde{\mathcal{K}}_k(\psi, \eta)d\mu(\psi)
$$

$$
= 2 \int_{(\frac{1}{2}, 1]} \tilde{\mathcal{K}}_k(\psi, \eta)d\mu(\psi)
$$

$$
= 2 \int_{(\frac{1}{2}, 1]} 2 \cdot \mathbf{1}_{(\frac{1}{2}, 1]}(\eta)d\mu(\psi) = 2 \cdot \mathbf{1}_{(\frac{1}{2}, 1]}(\eta).
$$

(ii) $\xi \in (2^{-(k+1)}, \frac{1}{2}]$: In this case, we have $\tilde{\mathcal{K}}_{k+1}(\xi, \psi) = 2^{k+1} \cdot \mathbf{1}_{[0, 2^{-(k+1)}]}(\psi)$ and hence,

$$
\tilde{\mathcal{K}}_{k+1}\tilde{\mathcal{K}}_k(\xi, \eta) = \int_{[0,1]} \tilde{\mathcal{K}}_{k+1}(\xi, \psi)\tilde{\mathcal{K}}_k(\psi, \eta)d\mu(\psi)
$$

$$
= 2^{k+1} \int_{[0, 2^{-(k+1)}]} \tilde{\mathcal{K}}_k(\psi, \eta)d\mu(\psi).
$$

Since $[0, 2^{-(k+1)}] \subset [0, 2^{-k}]$ it follows that,

$$
\tilde{\mathcal{K}}_{k+1}\tilde{\mathcal{K}}_k(\xi, \eta) = 2^{k+1} \int_{[0, 2^{-(k+1)}]} 2 \cdot \mathbf{1}_{(\frac{1}{2}, 1]}(\eta)d\mu(\psi) = 2 \cdot \mathbf{1}_{(\frac{1}{2}, 1]}(\eta).
$$

(iii) $\xi \in (\frac{1}{2}, 1]$: In this case, we have $\tilde{\mathcal{K}}_{k+1}(\xi, \psi) = 2 \cdot \mathbf{1}_{(\frac{1}{2}, 1]}(\psi)$ and hence, a similar result as in the case (i) holds.

Thus, $\tilde{\mathcal{K}}_{k+1}\tilde{\mathcal{K}}_k(\xi, \eta) = 2 \cdot \mathbf{1}_{(\frac{1}{2}, 1]}(\eta)$ implying that $\{\mathcal{K}_k\}$ is uniformly ergodic. Nevertheless, if we let $S = [0, \frac{1}{2}]$, then $\bar{\mathcal{K}}_k^f(S) = 2^{-k}$ and hence, $\sum_{k=0}^{\infty} \bar{\mathcal{K}}_k^f(S) = 2 < \infty$.

What makes the chain in Example 7.4 uniformly ergodic is the fact that the measure μ can approach zero and yet, such a small measure set can contribute a lot to ergodicity. As a consequence this straightforward generalization of the infinite flow property need not be necessary in general state spaces.

In fact, with proper definition of infinite flow property, the necessity of infinite flow is still true for arbitrary state spaces and for the weakest form of ergodicity, i.e. weak ergodicity. To formulate the infinite flow property, for a stochastic kernel $\mathcal{K}$ and any non-trivial set $S \in \mathcal{M}$, let us define the flow from set S to $\bar{S}$ to be:

$$
\mathcal{K}^f(S, \bar{S}) = \sup_{\xi \in S} \mathcal{K}(\xi, \bar{S}). \tag{7.7}
$$

Notice that the definition of the flow $\mathcal{K}^f(S, \bar{S})$ in (7.7) does not require the measurable space $(\mathcal{K}, \mathcal{M})$ to be equipped with a measure. Also, note that since $\mathcal{K}$ is a stochastic kernel, we have $\mathcal{K}^f(S, \bar{S}) \leq 1$.

Based on Eq. 7.7, let us define the flow between S and $\bar{S}$ to be $\mathcal{K}^f(S) = \mathcal{K}^f(S, \bar{S}) + \mathcal{K}^f(\bar{S}, S)$. Now, let us define the infinite flow property.

Definition 7.4 We say that a chain of stochastic kernels $\{\mathcal{K}_k\}$ has infinite flow property if $\sum_{k=0}^{\infty} \mathcal{K}_k^f(S) = \infty$ for any non-trivial set $S \in \mathcal{M}$.

For example, the chain $\{\mathcal{K}_k\}$ discussed in Example 7.4 has infinite flow property over the set $S = [0, \frac{1}{2}]$. In this case, we have $\mathcal{K}_k^f(S) = 1$ for any $k \geq 0$ and hence, $\sum_{k=0}^{\infty} \mathcal{K}_k^f([0, \frac{1}{2}]) = \infty$.

In fact, Definition 7.4 is a generalization of the infinite flow property in the classical averaging dynamics. In the case of a stochastic chain $\{\mathcal{K}_k\} = \{A(k)\}$, a non-trivial set $S \in \mathcal{M}$ would be a set with $S \neq \emptyset$ and $S \neq [m]$. For any such S, we have

$$\frac{1}{|S|}(A_{S\bar{S}}(k) + A_{\bar{S}S}(k)) \leq \mathcal{K}_k^f(S) \leq A_{S\bar{S}}(k) + A_{\bar{S}S}(k).$$

Thus, $\{A(k)\}$ has infinite flow property (in terms of Definition 7.4) if and only if

$$\sum_{k=0}^{\infty} A_{S\bar{S}}(k) + A_{\bar{S}S}(k) = \infty,$$

for any nonempty $S \subset [m]$ which coincides with Definition 3.2.

Now, we can prove the necessity of infinite flow property for the weak ergodicity.

Theorem 7.1 *The infinite flow property is necessary for weak ergodicity.*

Proof Let $\{\mathcal{K}_k\}$ be a chain that does not have infinite flow property. Therefore, there exists a non-trivial $S \in \mathcal{M}$ with $\sum_{k=0}^{\infty} \mathcal{K}_k^f(S) < \infty$. Let $t_0 \geq 0$ be such that $\sum_{t=t_0}^{\infty} \mathcal{K}_t^f(S) \leq \frac{1}{4}$. Let $x_{t_0} = \mathbf{1}_S - \mathbf{1}_{\bar{S}}$ which is in L_∞. Then, for any $k \geq t_0$, and any $\xi \in S$, we have

$$
\begin{aligned}
x_{k+1}(\xi) &= \int_X \mathcal{K}_k(\xi, d\eta) x_k(\eta) \\
&= \int_S \mathcal{K}_k(\xi, d\eta) x_k(\eta) + \int_{\bar{S}} \mathcal{K}_k(\xi, d\eta) x_k(\eta) \\
&\geq \inf_{\eta \in S}(x_k(\eta)) \mathcal{K}_k(\xi, S) - \mathcal{K}_k(\xi, \bar{S}) \\
&= \inf_{\eta \in S}(x_k(\eta))(1 - \mathcal{K}_k(\xi, \bar{S})) - \mathcal{K}_k(\xi, \bar{S}),
\end{aligned}
\tag{7.8}
$$

where the inequality follows from $\|x_k\|_\infty \leq \|x_{t_0}\|_\infty = 1$ (Lemma 7.1), and the last equality follows from $\mathcal{K}_k(\xi, \cdot)$ being a probability measure. Therefore, assuming $\inf_{\eta \in S}(x_k(\eta)) \geq 0$, we have

$$
\begin{aligned}
x_{k+1}(\xi) &\geq \inf_{\eta \in S}(x_k(\eta))(1 - \mathcal{K}_k(\xi, \bar{S})) - \mathcal{K}_k(\xi, \bar{S}) \\
&\geq \inf_{\eta \in S}(x_k(\eta))(1 - \mathcal{K}_k^f(S, \bar{S})) - \mathcal{K}_k^f(S, \bar{S}) \\
&\geq \inf_{\eta \in S}(x_k(\eta)) - 2\mathcal{K}_k^f(S, \bar{S}),
\end{aligned}
$$

where the last inequality follows by $\|x(k)\|_\infty \leq 1$ and the fact that $\mathcal{K}_k(\xi, \bar{S}) \leq \mathcal{K}_k^f(S, \bar{S})$. Therefore, using induction, for any $\xi \in S$, we can show that

$$\inf_{\eta \in S}(x_{k+1}(\eta)) \geq \inf_{\eta \in S}(x_k(\eta)) - 2\mathcal{K}_k^f(S) \geq \inf_{\eta \in S}(x_{t_0}(\eta)) - 2\sum_{t=t_0}^{k}\mathcal{K}_t^f(S) \geq \frac{1}{2},$$

and hence, $\inf_{\eta \in S}(x_k(\eta)) \geq \frac{1}{2} > 0$ for any $k \geq t_0$. Using the same line of argument, we have

$$\sup_{\eta \in \bar{S}} x_k(\eta) \leq \sup_{\eta \in \bar{S}}(x_{t_0}(\eta)) + 2\sum_{t=t_0}^{k}\mathcal{K}_t^f(S) \leq -\frac{1}{2}.$$

Therefore, for any $\xi \in S$ and any $\eta \in \bar{S}$, we have

$$\liminf_{k \to \infty}(x_k(\xi) - x_k(\eta)) \geq 1,$$

which shows that $\{\mathcal{K}_k\}$ is not weakly ergodic. **Q.E.D.**

Restriction of the proof of Theorem 7.1 to the stochastic chains in $\mathbb{R}^m$ gives an algebraic proof for Theorem 3.1.

7.4 Quadratic Comparison Function

In Chap. 4, we used a quadratic comparison function to perform stability analysis for averaging dynamics. For this, in Theorem 4.4, we derived an identity which quantifies the exact decrease rate for the introduced quadratic comparison function for deterministic chains. In this section, we show that the decrease rate provided in Theorem 4.4 can be generalized to an arbitrary state space.

Let π be a probability measure on $(X, \mathcal{M})$. For a stochastic kernel $\mathcal{K}$, let us define $\lambda : \mathcal{M} \to \mathbb{R}^+$ by $\lambda(S) = \int_X \pi(d\xi)\mathcal{K}(\xi, S)$ for any $S \in \mathcal{M}$. Then λ induces a measure on $(X, \mathcal{M})$ and also since $\lambda(X) = \int_X \pi(d\xi)\mathcal{K}(\xi, X) = \int_X \pi(d\xi) = 1$, the measure λ is a probability measure.

Motivated by the concept of absolute probability sequence for stochastic chains, let us have the following definition.

Definition 7.5 We say that a sequence of probability measures on $(X, \mathcal{M})$ is an absolute probability sequence for $\{\mathcal{K}_k\}$ if for any $S \in \mathcal{M}$, we have

$$\pi_k(S) = \int_X \pi_{k+1}(d\xi)\mathcal{K}_k(\xi, S) \qquad \text{for any } k \geq 0.$$

As in the case of classic averaging dynamics, using an absolute probability sequence and a convex function $g : \mathbb{R} \to \mathbb{R}$, we can construct a comparison function for dynamics driven by $\{\mathcal{K}_k\}$. For this, let us define $V_{g,\pi} : L_\infty \times \mathbb{Z}^+ \to \mathbb{R}^+$ by

$$V_{g,\pi}(x,k) = \int_X \pi_k(d\xi)g(x(\xi)) - g(\pi_k x) = \int_X \pi_k(d\xi)\left(g(x(\xi)) - g(\pi_k x)\right),$$

$$(7.9)$$

where $\{\pi_k\}$ is an absolute probability sequence for $\{\mathcal{K}_k\}$ and $\pi x = \int_X \pi(d\xi)x(\xi)$.

As for the classic averaging dynamics, for a probability measure π, we have

$$\int_X \pi(d\xi)\nabla g(\pi x)(x(\xi) - \pi x) = \nabla g(\pi x)\int_X \pi(d\xi)(x(\xi) - \pi x) = 0,$$

which follows from π being a probability measure.

Therefore, by subtracting $0 = \int_X \pi(d\xi)\nabla g(\pi_k x)(x(\xi) - \pi_k x)$ from the both sides of 7.9, we have

$$\begin{aligned}
V_{g,\pi}(x,k) &= \int_X \pi_k(d\xi)\left(g(x(\xi)) - g(\pi_k x)\right) \\
&= \int_X \pi_k(d\xi)\left(g(x(\xi)) - g(\pi_k x) - \nabla g(\pi_k x)(x(\xi) - \pi_k x)\right) \\
&= \int_X \pi_k(d\xi)D_g(x(\xi), \pi_k x),
\end{aligned}$$

where $D_g(x(\xi), \pi_k x)$ is the Bregman distance of $x(\xi)$ and $\pi_k x$ under further assumption of strong convexity of g. Thus, the given comparison function $V_{g,\pi}(x,k)$ is in fact, the weighted distance of a measurable function $x \in L_\infty$ from the average point $\pi_k x$.

The following result shows that $V_{g,\pi}(x,k)$ is a comparison function for the dynamics (7.2).

Theorem 7.2 *Let $\{\pi_k\}$ be an absolute probability sequence for $\{\mathcal{K}_k\}$. Then, $V_{g,\pi}(x,k)$, as defined in* Eq. 7.9, *is a comparison function for $\{\mathcal{K}_k\}$.*

Proof Let $\{x(k)\}$ be a dynamics driven by $\{\mathcal{K}_k\}$. Then, for any $k \geq t_0$, we have

$$\begin{aligned}
\int_X \pi_{k+1}(d\xi)g(x_{k+1}(\xi)) &= \int_X \pi_{k+1}(d\xi)g\left(\int_X \mathcal{K}_k(\xi, d\eta)x_k(\eta)\right) \\
&\leq \int_X \pi_{k+1}(d\xi)\int_X \mathcal{K}_k(\xi, d\eta)g(x_k(\eta)) \\
&= \int_X \pi_k(d\eta)g(x_k(\eta)), \qquad\qquad (7.10)
\end{aligned}$$

where the inequality follows by the application of the Jensen's inequality (Theorem A.2), and the last equality follows by $\{\pi_k\}$ being absolute probability sequence for $\{\mathcal{K}_k\}$. On the other hand, we have $\pi_{k+1}x_{k+1} = \pi_k x_k$ and hence, $g(\pi_{k+1}x_{k+1}) = g(\pi_k x_k)$. Using this observation and relation (7.10), we conclude that

$$\begin{aligned}
V_{g,\pi}(x_{k+1}, k+1) &= \int_X \pi_{k+1}(d\xi)g(x_{k+1}(\xi)) - g(\pi_{k+1}x_{k+1}) \\
&\leq \int_X \pi_k(d\xi)g(x_k(\xi)) - g(\pi_k x_k) \\
&= V_{g,\pi}(x_k, k).
\end{aligned}$$

Q.E.D.

Similar to the classical averaging dynamics, let us denote the quadratic comparison function $V_\pi : L_\infty \times \mathbb{Z}^+ \to [0, \infty)$ by

$$V_\pi(x, k) = \int_X \pi_k(d\xi)(x(\xi) - \pi_k x)^2 = \int_X \pi_k(d\xi)x^2(\xi) - (\pi_k x)^2. \qquad (7.11)$$

Note that for any stochastic function π and any $x \in L_\infty$, we have

$$\int_X \pi(d\xi)(x(\xi) - \pi x)^2 \leq 4\|x\|_\infty^2 < \infty,$$

which follows from π being stochastic and $x \in L_\infty$. Therefore, for any dynamics $\{x_k\}$ driven by a chain $\{\mathcal{K}_k\}$ and any absolute probability sequence $\{\pi_k\}$, we have

$$V_\pi(x_k, k) \leq 4\|x_k\|_\infty^2 \leq 4\|x_{t_0}\|_\infty^2,$$

which follows by Lemma 7.1.

Suppose that we are given a probability measure π on $(X, \mathcal{M})$ and a stochastic kernel $\mathcal{K}$. Our next step is to define a probability measure H on the product space $(X \times X, \mathcal{M} \otimes \mathcal{M})$ using π and $\mathcal{K}$. Let us define H on the product of measurable sets $S, T \in \mathcal{M}$ by

$$H(S \times T) = \int_X \pi(d\xi)\mathcal{K}(\xi, S)\mathcal{K}(\xi, T).$$

Moreover, for a collection of disjoint sets $S_1 \times T_1, \ldots, S_n \times T_n \subset X \times X$, where $S_i, T_i \in \mathcal{M}$ for all $i \in [n]$, let

$$H\left(\bigcup_{i=1}^n S_i \times T_i\right) = \sum_{i=1}^n H(S_i \times T_i) = \sum_{i=1}^n \int_X \pi(d\xi)\mathcal{K}(\xi, S_i)\mathcal{K}(\xi, T_i). \qquad (7.12)$$

Equation (7.12) provides a pre-measure on the algebra of rectangular sets on $X \times X$. By Theorem 1.14 in [2], H can be extended to an outer measure on $X \times X$ such that its restriction on $\mathcal{M} \otimes \mathcal{M}$ is a measure. Note that $H(X \times X) = \int_X \pi(d\xi)\mathcal{K}(\xi, X)\mathcal{K}(\xi, X) = 1$ and hence, H is a probability measure on the product space $(X \times X, \mathcal{M} \otimes \mathcal{M})$. We refer to the measure H constructed this way as the measure induced by $\mathcal{K}$ and π (on $(X \times X, \mathcal{M} \otimes \mathcal{M})$).

Now, we can quantify the decrease rate of the quadratic comparison function along any trajectory of the dynamics (7.2).

Theorem 7.3 *Suppose that π, λ are probability measures on $(X, \mathcal{M})$ such that $\lambda(S) = \int_X \pi(d\xi)\mathcal{K}(\xi, S)$ for all $S \in \mathcal{M}$. Then, for any x, $y \in L_\infty$ with $y = \mathcal{K}x$, we have*

$$\int_X \pi(d\xi)y^2(\xi) - (\pi y)^2 = \left\{ \int_X \lambda(d\xi)x^2(\xi) - (\lambda x)^2 \right\}$$
$$- \frac{1}{2} \int_{X \times X} H(d\eta_1 \times d\eta_2)(x(\eta_1) - x(\eta_2))^2, \tag{7.13}$$

where H is the probability measure induced by $\mathcal{K}$ and π on $(X \times X, \mathcal{M} \otimes \mathcal{M})$.

Proof We first prove relation (7.13) for arbitrary simple functions. Let $x = \sum_{i=1}^{m} \alpha_i \mathbf{1}_{S_i}$ where $\{S_i \mid i \in [m]\} \subseteq \mathcal{M}$ is a partition for X with $m \geq 1$ disjoint sets (i.e. $S_i \cap S_j = \emptyset$ (a.e.) for all $i, j \in [m]$ with $i \neq j$ and $\bigcup_{i \in [m]} S_i = X$) and also $\alpha = (\alpha_1, \dots, \alpha_m)^T \in \mathbb{R}^m$. Then, we have

$$y(\xi) = \mathcal{K}(\xi, \cdot)x = \int_X \mathcal{K}(\xi, d\eta)x(\eta) = \sum_{i=1}^{m} \alpha_i \mathcal{K}(\xi, S_i).$$

Therefore,

$$\pi y^2 = \int_X \pi(d\xi)y^2(\xi) = \int_X \pi(d\xi)\left(\sum_{i=1}^{m} \alpha_i \mathcal{K}(\xi, S_i)\right)^2$$
$$= \int_X \pi(d\xi)\left(\sum_{i=1}^{m} \alpha_i^2 \mathcal{K}_i^2(\xi, S_i)\right)$$
$$+ \int_X \pi(d\xi)\left(\sum_{\substack{i,j \in [m] \\ i \neq j}} \alpha_i \alpha_j \mathcal{K}(\xi, S_i)\mathcal{K}(\xi, S_j)\right). \tag{7.14}$$

By the linearity of integral we have

$$\int_X \pi(d\xi)\left(\sum_{\substack{i,j \in [m] \\ i \neq j}} \alpha_i \alpha_j \mathcal{K}(\xi, S_i)\mathcal{K}(\xi, S_j)\right)$$
$$= \sum_{\substack{i,j \in [m] \\ i \neq j}} \alpha_i \alpha_j \int_X \pi(d\xi)\mathcal{K}(\xi, S_i)\mathcal{K}(\xi, S_j) = \sum_{\substack{i,j \in [m] \\ i \neq j}} \alpha_i \alpha_j H(S_i \times S_j), \tag{7.15}$$

which follows by the definition of the induced measure H. On the other hand, for any $\xi \in X$ we have $\mathcal{K}(\xi, S_i) = 1 - \sum_{j \neq i} \mathcal{K}(\xi, S_j)$ which follows from $\mathcal{K}(\xi, \cdot)$ being a probability measure. Therefore,

$$\int_X \pi(d\xi) \left(\sum_{i=1}^m \alpha_i^2 \mathcal{K}^2(\xi, S_i) \right)$$

$$= \int_X \pi(d\xi) \left(\sum_{i=1}^m \alpha_i^2 \left\{ \mathcal{K}(\xi, S_i) - \sum_{j \neq i} \mathcal{K}(\xi, S_i)\mathcal{K}(\xi, S_j) \right\} \right).$$

We also have

$$\int_X \pi(d\xi) \left(\sum_{i=1}^m \alpha_i^2 \mathcal{K}(\xi, S_i) \right) = \sum_{i=1}^m \alpha_i^2 \left(\int_X \pi(d\xi)\mathcal{K}(\xi, S_i) \right)$$

$$= \sum_{i=1}^m \alpha_i^2 \lambda(S_i) = \lambda x^2, \tag{7.16}$$

which follows from the definition of the probability measure λ and $x = \sum_{i=1}^m \alpha_i \mathbf{1}_{S_i}$. Also, using a similar argument as the one used to derive Eq. 7.15, we have

$$\int_X \pi(d\xi) \left(\sum_{i=1}^m \alpha_i^2 \sum_{j \neq i} \mathcal{K}(\xi, S_i)\mathcal{K}(\xi, S_j) \right) = \sum_{\substack{i,j \in [m] \\ i \neq j}} H(S_i \times S_j)\alpha_i^2. \tag{7.17}$$

Therefore, replacing relations (7.15), (7.16), and (7.17) in relation (7.14), we have:

$$\pi y^2 = \lambda x^2 - \left\{ \sum_{\substack{i,j \in [m] \\ i \neq j}} H(S_i \times S_j)(\alpha_i^2 - \alpha_i \alpha_j) \right\}$$

$$= \lambda x^2 - \frac{1}{2} \sum_{i,j \in [m]} H(S_i \times S_j)(\alpha_i - \alpha_j)^2,$$

where the last equation holds since, $H(S_i \times S_j) = H(S_j \times S_i)$ for any $i, j \in [m]$. Note that the function $x_i(\eta_1) - x_j(\eta_2)$ is equal to the constant $\alpha_i - \alpha_j$ over $S_i \times S_j$. Thus,

$$\sum_{i,j \in [m]} H(S_i \times S_j)(\alpha_i - \alpha_j)^2 = \int_{X \times X} H(d\eta_1 \times d\eta_2)(x(\eta_1) - x(\eta_2))^2. \tag{7.18}$$

The last step to prove the result for simple functions is to use that fact that $(\pi y)^2 = (\pi(\mathcal{K}x))^2 = (\lambda x)^2$. Therefore, by subtracting this relation from the both sides of (7.18) we arrive at the desired relation (7.13).

Now, we prove that the assertion holds for an arbitrary $x \in L_\infty$. Note that the function $T(z) = \pi z = \int_X \pi(d\xi)z(\xi)$ is a continuous function on $(L_\infty, \|\cdot\|_\infty)$ for any probability measure π. Also, $\tilde{T}(x) = x^2$ is a continuous function from $L_\infty \to L_\infty$.

Thus the function $\pi y^2 - (\pi y)^2$ and $\lambda x^2 - (\lambda x)^2$ are continuous functionals over L_∞. Similarly, the functional $\bar{T}(x) = \int_{X \times X} H(d\eta_1 \times d\eta_2)(x(\eta_1) - x(\eta_2))^2$ is a continuous functional from $(L_\infty, \|\cdot\|_\infty)$ to $\mathbb{R}$, which follows from H being a probability measure on $(X \times X, \mathcal{M} \otimes \mathcal{M})$. Therefore, all the functionals involved in relation (7.13) are continuous over L_∞. Since the relation holds for a dense subset of L_∞ (i.e. simple functions), we conclude that the relation holds for any $x \in L_\infty$.   **Q.E.D**

Using Theorem 7.3, the following corollary follows immediately.

Corollary 7.1 *Let $\{\mathcal{K}_k\}$ be a chain of stochastic integral kernels and let $\{\pi_k\}$ be an absolute probability sequence for $\{\mathcal{K}_k\}$. Then, for any dynamics $\{x_k\}$ driven by $\{\mathcal{K}_k\}$ started at time t_0 and point $x_{t_0} \in L_\infty$, we have:*

$$V_\pi(x_{k+1}, k+1) = V_\pi(x_k) - \frac{1}{2} \int_{X \times X} H_k(d\eta_1 \times d\eta_2) \, (x_k(\eta_1) - x_k(\eta_2))^2 \, ,$$

where H_k is the induced measure on $(X \times X, \mathcal{M} \otimes \mathcal{M})$ by $\mathcal{K}_k$ and π_{k+1}. Furthermore, we have

$$\sum_{k=t_0}^{\infty} \int_{X \times X} H_k(d\eta_1 \times d\eta_2) \, (x_k(\eta_1) - x_k(\eta_2))^2 \leq 2V_\pi(x_{t_0}, t_0).$$

References

1. E. Nummelin, *General Irreducible Markov Chains and Non-Negative Operators* (Cambridge University Press, Cambridge, 1984)
2. G.B. Folland, *Real Analysis: Modern Techniques and Their Applications*, 2nd edn. (Wiley, New York, 1999)

Chapter 8
Conclusion and Suggestions for Future Works

8.1 Conclusion

We studied products of random stochastic matrices and the limiting behavior of averaging dynamics as well as averaging dynamics over arbitrary state spaces. The main idea and contribution of this thesis is to demonstrate that the study of the limits of product of stochastic matrices is closely related to the study of the summation of stochastic matrices, i.e. the study of *flows*.

We introduced the notions of infinite flow property, absolute infinite flow property, infinite flow graph, and infinite flow stability for a chain of stochastic matrices and showed that they are closely related to the limiting behavior of product of stochastic matrices and averaging dynamics. We proved that for a class of stochastic matrices, i.e. the class $\mathcal{P}^*$ with feedback property, this limiting behavior can be determined by investigating the infinite flow graph of the given chain. Our proof is based on the use of a quadratic comparison function and also the derived decrease rate for such a comparison function along any trajectory of the averaging dynamics. We defined balanced property for a chain of stochastic matrices which can be verified efficiently. We proved that any balanced chain with feedback property is an instance of a class $\mathcal{P}^*$ chain with feedback property and hence, the product of such stochastic matrices are convergent and the structure of the limiting matrices can be determined using the infinite flow graph of the chain. We showed that this class contains many of the previously studied chains of stochastic matrices.

We then studied the implications of the developed results for independent random chains, uniformly bounded chains, product of inhomogeneous stochastic matrices, link-failure models and Hegselmann-Krause model for opinion dynamics. We also provided an alternative proof of the second Borel-Cantelli lemma.

Inspired by the necessity of infinite flow property for ergodicity of stochastic chains, we proved that a stronger property, the absolute infinite flow property, is necessary for ergodicity. We showed that, in fact, this property is equivalent to ergodicity for doubly stochastic chains. To prove these results we introduced the rotational transformation of a stochastic chain with respect to a permutation chain and showed that

B. Touri, *Product of Random Stochastic Matrices and Distributed Averaging*,
Springer Theses, DOI: 10.1007/978-3-642-28003-0_8,
© Springer-Verlag Berlin Heidelberg 2012

many of the limiting behavior of the product of stochastic matrices is invariant under rotational transformation.

Finally, we generalized the framework for the study of the averaging dynamics over general state spaces. We defined several modes of ergodicity and showed that unlike the averaging dynamics over $\mathbb{R}^m$, different modes of ergodicity are not equivalent in general state spaces. Then, we introduced a generalization of the infinite flow property to arbitrary state spaces and showed that this generalization remains to be necessary for the weakest form of ergodicity in arbitrary state spaces. We also introduced the concept of absolute infinite flow property for a chain of stochastic integral kernel. We showed that, as in the case of averaging dynamics in finite state spaces, using an absolute probability sequence for a chain of stochastic integral kernel, one can develop infinitely many comparison function for any dynamics driven by the chain. Moreover, we quantified the decrease rate of the quadratic comparison function along the trajectory of the averaging dynamics.

8.2 Suggestions for Future Works

Here, we discuss some questions and suggestions for future works on product of random stochastic matrices and weighted averaging dynamics.

Chapter 3

1. One can define the *directed* infinite flow graph $\bar{G}^\infty = ([m], \bar{\mathcal{E}}^\infty)$ by letting

$$\bar{\mathcal{E}}^\infty = \{(i, j) \mid i \neq j, \sum_{k=0}^{\infty} A_{ij}(k) = \infty\},$$

   which contains more information than the infinite flow graph. What properties of the limiting behavior of the averaging dynamics can be deduced from the directed infinite flow graph which cannot be deduced from the infinite flow graph itself?

Chapter 4

1. As it is shown in Example 4.1, there is a gap between the infinite flow stability and convergence of the product $A(k : t_0)$ for any $t_0 \geq 0$. Characterization of the chains that have the latter property but they are not infinite flow stable remains open.
2. Does Theorem 4.10 hold for adapted chains that are balanced in expectation? As it is shown in Chap. 5, a lower bound $p^* > 0$ for coordinates of an absolute probability sequence plays an important role on development of an upper bound for the convergence rate of averaging dynamics. For balanced chains with feedback property, we have provided a lower bound p^* which depends on the minimum value of the positive entries of matrices in the polyhedral set $\mathbf{B}_{\alpha,\beta}$ (defined in Eq. (4.18)). Some related open questions include:

    I. Can the bound $\min(\frac{1}{m}, \gamma^{m-1})$ in Corollary 4.6 be improved?

    II. What is the characterization of the extreme points in the polyhedral set $\mathbf{B}_{\alpha,\beta}$?

    III. What is the minimum value for the positive entries of the matrices in $\mathbf{B}_{\alpha,\beta}$?

Chapter 5

1. Theorem 5.2 and Theorem 5.5 provide upper bounds for the convergence rate of averaging dynamics. Derivation of a lower bound for the convergence rate of averaging dynamics is left open for future work.
2. The developed machinery in Chaps. 3, 4, and 5 suggests a way of extending the celebrated Cheeger's inequality to time-inhomogeneous chains. Such an extension remained open for future works.
3. For the link-failure models investigated in Sect. 5.4, one can use the same machinery to study *link attenuation* models. However, we did not find any practical scenario that such a model would fit into it.
4. The study of link failure models for adapted processes remains open for future work.
5. Can the generic lower bound $p^* \geq \frac{1}{m^{m-1}}$ for entries of an absolute probability sequence be customized and improved for Hegselmann-Krause dynamics?

Chapter 6

1. We showed that a doubly stochastic chain is ergodic if and only if it has absolute infinite flow property. For what other classes of stochastic chains does this equivalency hold?

Chapter 7

There are many natural questions that can be asked related to the content of this chapter. This chapter was intended to show that the study of the averaging dynamics in $\mathbb{R}^m$ can be generalized to arbitrary state spaces. Here are few immediate questions that would be of interest:

1. What is a proper way of generalizing B-connectivity to general state spaces?
2. What is a proper way of defining the infinite flow graph for a chain of stochastic kernels?
3. For a finite measure space X, i.e. $\mu(X) < \infty$, one can define a chain of stochastic kernels $\{\mathcal{K}_k\}$ to be doubly stochastic if $\{\frac{1}{\mu(X)} 1_X\}$ is an absolute probability sequence for $\{\mathcal{K}_k\}$. With a proper definition of feedback property, is the infinite flow property sufficient for weak ergodicity of a chain of doubly stochastic kernels with feedback property?
4. Under what condition an absolute probability sequence exists for a chain of stochastic kernels?

5. One can extend the definition of balancedness for a chain of stochastic kernels $\{\mathcal{K}_k\}$ by requiring that $\mathcal{K}_k^f(S, \bar{S}) \geq \alpha \mathcal{K}_k^f(\bar{S}, S)$ for some $\alpha > 0$, any non-trivial $S \in \mathcal{M}$ and any $k \geq 0$. Can we show results such as Theorem 4.10 (with a proper definition of feedback property) for a chain of balanced stochastic kernels?

Appendix A
Background Material on Probability Theory

Here, we review some of the results and tools from probability theory. Materials of this section are extracted from [1]. We assume that readers are familiar with basic notions of probability theory such as probability space, expectation, and conditional expectation. We first review some results on the convergence of sequences of random variables. Next, we present some results on conditional expectation and the martingale convergence theorem. Finally, we discuss some results on sequences of independent random chains.

A.1 Convergence of Random Variables

Let Ω be a sample space and $\mathcal{F}$ be a σ-algebra of subsets of Ω. Let $\Pr(\cdot)$ be a probability measure on $(\Omega, \mathcal{F}, \Pr(\cdot))$. Many of the events of our interest can be formulated as a set, or intersection of sets that are described by a limit of a sequence of random variables. To be able to discuss probability of such sets, we should be assured that such sets are in fact measurable sets. The following result gives such a certificate.

Lemma A.1 *For a sequence of random variables $\{u(k)\}$, the limits $\limsup_{k\to\infty} u(k)$ and $\liminf_{k\to\infty} u(k)$ are measurable. In particular, the sets $\{\omega \mid \lim_{k\to\infty} u(k)\ exists\}$ and $\{\omega \mid \lim_{k\to\infty} u(k) = \alpha\}$ are measurable sets for any $\alpha \in [-\infty, \infty]$.*

Lemma A.1 follows from Theorem 2.5 in [1]. As an immediate corollary to Lemma A.1, the following result holds.

Corollary A.1 *For a sequence of random vectors $\{x(k)\}$, the set $\{\omega \mid \lim_{k\to\infty} x(k)\ exists\}$ is a measurable set.*

B. Touri, *Product of Random Stochastic Matrices and Distributed Averaging*, Springer Theses, DOI: 10.1007/978-3-642-28003-0, © Springer-Verlag Berlin Heidelberg 2012

Proof Note that $\{\omega \mid \lim_{k\to\infty} x(k) \text{ exists}\} = \bigcap_{i=1}^{m}\{\omega \mid \lim_{k\to\infty} x_i(k) \text{ exists}\}$ and by Lemma A.1, the set $\{\omega \mid \lim_{k\to\infty} x_i(k) \text{ exists}\}$ is measurable for all $i \in [m]$. Since, $\mathcal{F}$ is a σ-algebra, thus we conclude that $\{\omega \mid \lim_{k\to\infty} x(k,\omega) \text{ exists}\}$ is measurable.   **Q.E.D.**

For a sequence $\{u(k)\}$ of random variables that is convergent almost surely, we are often interested in conditions under which we can swap limit and expectation, i.e. $\lim_{k\to\infty} \mathsf{E}[u(k)] = \mathsf{E}[\lim_{k\to\infty} u(k)]$. The following two results give us two conditions under which we can interchange the limit and expectation operations.

Theorem A.1 (Monotone Convergence theorem) *Suppose that $\{u(k)\}$ is a sequence of non-decreasing and non-negative random variables, i.e. $0 \leq u(k)(\omega) \leq u(k+1)(\omega)$ for all $\omega \in \Omega$. Then,*

$$\lim_{k\to\infty} \mathsf{E}[u(k)] = \mathsf{E}\left[\lim_{k\to\infty} u(k)\right].$$

Theorem A.2 (Dominated Convergence theorem) *Suppose that $\{u(k)\}$ is a sequence of random variables dominated by a non-negative random variable v, i.e. $|u(k)| \leq v$ almost surely. Then if $\mathsf{E}[v < \infty]$, we have*

$$\lim_{k\to\infty} \mathsf{E}[u(k)] = \mathsf{E}\left[\lim_{k\to\infty} u(k)\right].$$

Proofs for Theorems A.1 and A.2 can be found in [1], p. 15.

As a consequence of the monotone convergence theorem, we have the following result.

Corollary A.2 *Let $\{u(k)\}$ be a sequence of non-negative random variables. Then,*

$$\mathsf{E}\left[\sum_{k=0}^{\infty} u(k)\right] = \sum_{k=0}^{\infty} \mathsf{E}[u(k)].$$

Proof Let $\{s(k)\}$ be the sequence of partial sums of $\{u(k)\}$, i.e. $s(k) = \sum_{t=0}^{k} u(t)$. Since $u(t)$s are non-negative, $\{s(k)\}$ is a non-decreasing and non-negative sequence and hence, the result follows by the monotone convergence theorem and linearity of the expectation operation.   **Q.E.D.**

A.2 Conditional Expectation and Martingales

Here, we review some of the results on conditional expectation and, also, present the martingale convergence theorem.

We start by presenting the Jensen's inequality ([1], p. 223).

Lemma A.2 (Jensen's Inequality) *Let $\phi : \mathbb{R} \to \mathbb{R}$ be a convex function and let u be a random variable. Then, for any sub σ-algebra $\tilde{\mathcal{F}} \subseteq \mathcal{F}$, we have $\phi(\mathsf{E}[u \mid \tilde{\mathcal{F}}]) \leq \mathsf{E}[\phi(u) \mid \tilde{\mathcal{F}}]$.*

Another result that is useful in our development is the following ([1], p. 224).

Lemma A.3 ([1], p. 224) *Consider two sub σ-algebras $\mathcal{F}_1 \subseteq \mathcal{F}_2 \subseteq \mathcal{F}$. Then, for any random variable u, we have*

$$\mathsf{E}[\mathsf{E}[u \mid \mathcal{F}_1] \mid \mathcal{F}_2] = \mathsf{E}[\mathsf{E}[u \mid \mathcal{F}_2] \mid \mathcal{F}_1] = \mathsf{E}[u \mid \mathcal{F}_1].$$

Suppose that we have a sequence $\{\mathcal{F}_k\}$ of σ-algebras in $(\Omega, \mathcal{F})$ such that $\mathcal{F}_k \subseteq \mathcal{F}$ and $\mathcal{F}_k \subseteq \mathcal{F}_{k+1}$ for all $k \geq 0$. Such a sequence of σ-algebras is referred to as a *filtration*. A sequence of random variables $\{u(k)\}$ is said to be adapted to $\{\mathcal{F}_k\}$ if $u(k)$ is measurable with respect to $\mathcal{F}_k$ for any $k \geq 0$.

An adapted sequence $\{u(k)\}$ of random variables is said to be a martingale with respect to a filtration $\{\mathcal{F}_k\}$, if we

- $\{u(k)\}$ is adapted to $\{\mathcal{F}_k\}$,
- $\mathsf{E}[|u(k)|] < \infty$ for all $k \geq 0$, and
- $\mathsf{E}[u(k+1) \mid \mathcal{F}_k] = u(k)$ for all $k \geq 0$,

and it is said to be a super-martingale, if $\mathsf{E}[u(k+1) \mid \mathcal{F}_k] \leq u(k)$ for all $k \geq 0$.

Using Jensen's equality which is provided in Lemma A.2, we conclude the following result.

Lemma A.4 ([1], p. 230) *Let $\phi : \mathbb{R} \to \mathbb{R}$ be a convex function and let $\{u(k)\}$ be a martingale adapted to a filtration $\{\mathcal{F}_k\}$. Then, $\{-\phi(u(k))\}$ is a super-martingale sequence with respect to $\{\mathcal{F}_k\}$.*

Proof For any $k \geq 0$, we have

$$\mathsf{E}[-\phi(u(k+1)) \mid \mathcal{F}_k] \leq -\phi(\mathsf{E}[u(k+1) \mid \mathcal{F}_k]) = -\phi(u(k)),$$

where the inequality follows by Lemma A.2 and the equality follows by the fact that $\{u(k)\}$ is a martingale sequence.   **Q.E.D.**

The following result shows that any super-martingale (and hence, martingale) which is bounded below, in a certain way, is convergent almost surely.

Theorem A.3 (Martingale Convergence theorem [1], p. 233) *Let $\{u(k)\}$ be a super-martingale sequence. Also, suppose that $\sup_{k \geq 0} \mathsf{E}[\max(-u(k), 0)] < \infty$. Then, $\{u(k)\}$ is convergent almost surely.*

Note that if $\{u(k)\}$ is a non-negative super-martingale, then we have $\max(-u(k), 0) = 0$, and hence, any non-negative super-martingale sequence is convergent almost surely.

Corollary A.3 *Let $\{u(k)\}$ be a sequence of non-negative super-martingale. Then, $\{u(k)\}$ is a convergent sequence almost surely.*

We will use this simplified version of the martingale convergence for our study.

Another important martingale result, which is often used to prove convergence of random sums, is the following result.

Theorem A.4 ([2], *p.164*) *Let $\{u(k)\}$, $\{\alpha(k)\}$, $\{\beta(k)\}$, and $\{\xi(k)\}$ be sequences of adapted non-negative random variables such that almost surely, for all $k \geq 0$,*

$$\mathsf{E}[u(k+1) \mid \mathcal{F}_k] \leq (1 + \alpha(k))u(k) + \beta(k) - \xi(k),$$

where $\sum_{k=0}^{\infty} \alpha(k) < \infty$ and $\sum_{k=0}^{\infty} \beta(k) < \infty$ almost surely. Then, $\lim_{k \to \infty} u(k)$ exists and $\sum_{k=0}^{\infty} \xi(k) < \infty$ almost surely.

A.3 Independence

The independency for a sequence of random variables, vectors, or matrices, allows us to use powerful tools that are available for independent sequences. Here, we discuss three of those tools that are used in this thesis.

A.3.1 Borel–Cantelli Lemma

Suppose that we are given a sequence of events $\{E(k)\}$ in a probability space $(\Omega, \mathcal{F}, \mathsf{Pr}(\cdot))$. The *infinitely often* (abbreviated as i.o.) event associated with $\{E(k)\}$ is defined by

$$\{E(k) \text{ i.o.}\} = \bigcap_{k=0}^{\infty} \bigcup_{t=k}^{\infty} E(t).$$

In other words, $\{E(k) \text{ i.o.}\}$ consists of the sample points $\omega \in \Omega$ that will occur in infinitely many of the events $\{E(k)\}$.

First and second Borel–Cantelli lemmas relate the $\sum_{k=0}^{\infty} \mathsf{Pr}(E(k))$ and the probability $\mathsf{Pr}(\{E(k) \text{ i.o.}\})$.

Lemma A.5 (First Borel–Cantelli lemma) *Let $\{E(k)\}$ be a sequence of (not necessarily independent) events on $(\Omega, \mathcal{F})$. Then, $\mathsf{Pr}(\{E(k)\} \, i.o.) > 0$ implies $\sum_{k=0}^{\infty} \mathsf{Pr}(E(k)) = \infty$.*

The second Borel–Cantelli lemma, provides the converse of this result given the independency of $\{E(k)\}$.

Lemma A.6 (Second Borel–Cantelli lemma) *Let $\{E(k)\}$ be a sequence of independent events on $(\Omega, \mathcal{F})$. Then, $\sum_{k=0}^{\infty} \mathsf{Pr}\,(E(k)) = \infty$ implies $\mathsf{Pr}\,(\{E(k)\}i.o.)$ $= 1$.*

Proofs for Lemmas A.5 and A.6 can be found in [1], pp. 46 and 49, respectively.

A.3.2 Kolmogorov's 0-1 Law

Consider a sequence $\{\mathcal{F}_k\}$ of σ-algebras. Consider the σ-algebra of the tale events of $\{\mathcal{F}_k\}$, i.e.

$$\tau = \bigcap_{k=0}^{\infty} \sigma\left(\bigcup_{t=k}^{\infty} \mathcal{F}_t\right).$$

As an example, let $\{u(k)\}$ be a scalar sequence adapted to $\{\mathcal{F}_k\}$ and let E to be the event that $\sum_{k=0}^{\infty} u(k)$ is convergent, i.e. $E = \{\omega \mid \sum_{k=0}^{\infty} u(k) \text{ is convergent}\}$. For a sequence $\{a(k)\}$ of real numbers, $\sum_{k=0}^{\infty} a(k)$ is convergent if and only if $\sum_{t=k}^{\infty} a(t)$ is convergent for any $k \geq 0$. Thus, E belongs to the tale of $\{\mathcal{F}_k\}$.

Now, we are ready to assert the Kolmogorov's 0-1 law.

Theorem A.5 (Kolmogorov's 0-1 law [1], p. 61) *Let $\{\mathcal{F}_k\}$ be a sequence of mutually independent σ-algerbas. Then, any tale event is a trivial event, i.e. it happens with either probability one or probability zero.*

A.3.3 Kolmogorov's Three-Series Theorem

Suppose that we have a sequence of independent random variables $\{u(k)\}$. The following result enables us to study convergence of random series $\sum_{k=0}^{\infty} u(k)$ using convergence of deterministic series.

Theorem A.6 (Kolmogorov's three-series theorem [1], p. 63) *Suppose that $\{u(k)\}$ is a sequence of independent random variables. Then, $\sum_{k=0}^{\infty} u(k)$ is convergent almost surely, if and only if, for any $\alpha > 0$ we have*

   I. $\sum_{k=0}^{\infty} \mathsf{Pr}\,(|u_n| > \alpha) < \infty$,
  II. $\sum_{k=0}^{\infty} \mathsf{E}\,[u(k)\mathbf{1}_{|u(k)| \leq \alpha}]$ *is convergent, and*
 III. $\sum_{k=0}^{\infty} var(u(k)\mathbf{1}_{|u(k)| \leq \alpha})$ *is convergent.*

As a consequence of Kolmogorov's three series theorem, consider the case where we have a sequence $\{u(k)\}$ of random variables in [0, 1]. Then, for $\alpha = 1$ we have $u(k)\mathbf{1}_{|u(k)| \leq \alpha} = u(k)$ and also $\Pr(|u_n| > \alpha) = 0$. On the other hand, since $u(k)$ is in [0, 1], we have $\mathsf{E}[u^2(k)] \leq \mathsf{E}[u(k)]$ and hence, $0 \leq var(u(k)) \leq \mathsf{E}[u(k)]$. Therefore, the following result is true.

Corollary A.4 *Let $\{u(k)\}$ be a sequence of independent random variables in* [0, 1]. *Then,* $\sum_{k=0}^{\infty} u(k) < \infty$ *almost surely if and only if* $\sum_{k=0}^{\infty} \mathsf{E}[u(k)] < \infty$.

Appendix B
Background Material on Real Analysis and Topology

Here, we discuss some background material on functional analysis and point set topology.

B.1 Functional Analysis

Consider a measurable space $(X, \mathcal{M}, \mu)$, where X is a set, $\mathcal{M}$ is a σ-algebra on X, and μ is a measure on $(X, \mathcal{M})$. For $p \in [1, \infty)$, consider the space L_p with the norm defined by $\|f\|_p = \left(\int_X |f|^p d\mu \right)^{1/p}$ and L_∞ with the norm $\|f\|_\infty = \mathrm{ess.}\ \sup(f)$ where $\mathrm{ess.}\ \sup(f)$ is the essential supremum of the function f. We denote the ball of radius r of the L_p space by $\mathcal{B}_p(r)$, i.e. $\mathcal{B}_p(r) = \{f \in L_p \mid \|f\|_p \leq r\}$.

Theorem B.1 (Hölder's inequality [3], p. 182) *For any $p, q \in [1, \infty]$ with $\frac{1}{p} + \frac{1}{q} = 1$, and for any $f \in L_p$ and $g \in L_q$, we have:*

$$\int_X |fg| d\mu \leq \|f\|_p \|g\|_q.$$

We say that a function f is a simple function if $f = \sum_{i=1}^{n} \alpha_i \mathbf{1}_{E_i}$ for some $E_1, \ldots, E_n \in \mathcal{M}$. Then, we have

Lemma B.1 *The set of simple functions is dense in L_∞.*

B. Touri, *Product of Random Stochastic Matrices and Distributed Averaging*,
Springer Theses, DOI: 10.1007/978-3-642-28003-0,
© Springer-Verlag Berlin Heidelberg 2012

B.2 Point Set Topology

Here, we mention two results that are used to provide an alternative proof of the existence of an absolute probability sequence for arbitrary stochastic chains in Chap. 4.

Theorem B.2 (Tychonoff theorem [4], p. 234) *An arbitrary product of compact spaces is compact in the product topology.*

Another result which is also due to Tychonoff, is the Tychonoff Fixed Point theorem.

Theorem B.3 (Tychonoff Fixed Point theorem [5], p. 47) *Let X be a non-empty compact subset of a locally convex linear topological space. Then, any continuous map $T : X \to X$ admits a fixed point, i.e. there exists $x \in X$ such that $T(x) = x$.*

References

1. R. Durrett, Probability: *Theory and Examples*, 3rd ed. (Curt Hinrichs, 2005)
2. A.S. Poznyak, *Advanced Mathematical Tools for Automatic Control Engineers: Stochastic Techniques* (Elsevier, Amsterdam, 2009)
3. G.B. Folland, *Real Analysis: Modern Techniques and Their Applications*, 2nd edn. (Wiley, London, 1999)
4. J. Munkres, *Topology*, 2nd ed. (Prentice Hall, Englewood Cliffs)
5. R. Agarwal, D. O'Regan, D. Sahu, *Fixed Point Theory for Lipschitzian-type Mappings with Applications*, vol. 150 (Springer, Berlin)

About the Author

Behrouz Touri received the B.S. degree in electrical engineering from Isfahan University of Technology, Isfahan, Iran, in 2006, the M.S. degree in communications, systems, and electronics from Jacobs University, Bremen, Germany, in 2008, and his Ph.D. in Industrial Engineering from the Department of Industrial and Enterprise Systems Engineering, University of Illinois at Urbana-Champaign in 2011. He is currently a postdoctoral research associate with the Coordinated Science Laboratory of the University of Illinois. His research interests include learning and dynamics over random networks, stability theory of

random and deterministic dynamics, decentralized optimization, control and estimation, and coding theory. He is a finalist for the best student paper award of the Conference on Decision and Control 2010, and the third place prize winner of the 13th International Mathematics Competition for university students, 2006.

B. Touri, *Product of Random Stochastic Matrices and Distributed Averaging*, Springer Theses, DOI: 10.1007/978-3-642-28003-0,
© Springer-Verlag Berlin Heidelberg 2012

Index

A

Absolute probability process, 42
Absolute probability sequence, 40
    general state space, 118
Adapted random chain, 23
Aperiodicity, 76
    inhomogenous chains, 4
Approximation lemma, 30

B

B-connected chain, 19
Balanced chain, 55
    independent chains, 58
Birkhoff-von Neumann theorem, 20
Borel-Cantelli lemma
    first lemma, 20
    second lemma, 67

C

Class $\mathcal{P}^*$, 49
Common steady state π, 57
Comparison function, 11
Consensus, 17
    general state space, 114
Consensus subspace, 18

D

Dominated convergence theorem, 132
Doubly stochastic matrix, 2

E

Ergodic index, 29
Ergodicity, 15
    strong, 15
    weak, 15
    general state space, 114

F

Feedback property, 37
    strong, 37
    weak, 37
Final component, 85
Flow in general state spaces, 119

G

Graph, 10
    connected, 10
    connected component, 10
    directed, 10
    strongly connected, 10
    undirected, 10

H

Hölder's inequality, 137

I

Infinite flow graph, 29
    for random chains, 34

B. Touri, *Product of Random Stochastic Matrices and Distributed Averaging,* 141
Springer Theses, DOI: 10.1007/978-3-642-28003-0,
© Springer-Verlag Berlin Heidelberg 2012

I (*cont.*)
Infinite flow property, 25
    general state spaces, 119
Infinite flow stability, 36
Inhomogeneous chain, 4
Irreducibility, 76
    inhomogeneous chains, 78

J
Jensen's inequality, 132

K
Kolmogorov's 0-1 law, 135
Kolmogorov's three-series
        theorem, 135

L
ℓ_1-approximation, 30
Lyapunov function, 11

M
Martingale convergence theorem, 133
Monotone convergence theorem, 132
Mutual ergodicity, 30
    random chains, 34

P
Permutation matrix, 9

R
Random chain, 23
    adapted, 24
    i.i.d., 24
    independent, 24

S
Static chain, 33
Stochastic matrix, 33

T
Termination time, 83
    for a final component, 83
Tychonoff Fixed Point theorem, 138
Tychonoff theorem, 137

U
Uniformly bounded chain, 78